READINGS ON

LAWS OF NATURE

READINGS on Laws of Nature

EDITED BY
JOHN W. CARROLL

UNIVERSITY OF
PITTSBURGH PRESS

Published by the University of Pittsburgh Press,
Pittsburgh, Pa. 15260

Manufactured in the United States of America
Printed on acid-free paper

10 9 8 7 6 5 4 3 2 1

Library of Congress Cataloging-in-Publication Data
Carroll, John W.
Readings on laws of nature / edited by John W. Carroll.
p. cm.
Includes bibliographical references and index.
ISBN 0-8229-5852-X (alk. paper)
1. Science—Philosophy. I. Carroll, John W. II. Title.
Q175.C36 2004
501—dc22
2004003482

CONTENTS

PREFACE

Serious contemporary philosophical discussion of laws of nature began in the late 1940s. Still, for most of the past fifty years or so, the topic has been addressed by philosophers whose primary interests were elsewhere. Philosophers analyzed laws because they believed them to be one of the keys to understanding the nature of scientific explanation, the truth conditions of counterfactuals, and the nature of induction. David Armstrong's 1983 *What Is a Law of Nature?* was the first clear sign that the problem of laws had become a topic of independent interest. Following Armstrong's book, there have been other signs: Bas van Fraassen's *Laws and Symmetries,* my own *Laws of Nature,* and Marc Lange's *Natural Laws in Scientific Practice.* As the topic has gradually come into its own, many philosophers have written seminal journal articles on laws. My goal for *Readings on Laws of Nature* was to gather the most central and accessible of these papers.

Keeping to my goal did have some unfortunate consequences, the most important one being that I did not include selections from either David Armstrong or David Lewis. I did not happily make that decision but believe it was appropriate. Armstrong has had enormous influence through his book; furthermore, his universals approach has significant similarities with the views advanced here by Fred Dretske and Michael Tooley. As for Lewis, his most weighty work on laws is scattered among several journal articles and books. Rather than trying to piece something together, it seemed best to let a version of his systems approach be defended here in the selection by Barry Loewer. Both Armstrong's and Lewis's views are reported and discussed in my introduction and several of the readings. To state it another way, Armstrong and Lewis have been so influential that there is no need for their work to appear in this collection!

Editing has been minimal. The papers have been left fairly unscathed except for some regularization of format.

Thank you to the contributing authors and the University of Pittsburgh Press for their support of this project. Thanks also to Ann Rives for her remarkably efficient off-hour scanning and preliminary proofreading. Doug Jesseph always had quick sensible answers to what struck me as difficult questions. Several of the authors (but especially Marc Lange, Barry Loewer, and John Roberts), beside providing their wisdom in the medium of their papers, had extremely helpful advice about the anthology. The support and love of my family has been unfailing.

READINGS ON

LAWS OF NATURE

Introduction

JOHN W. CARROLL

Four issues are central to this collection. First and foremost, there is the puzzling distinction between laws of nature and accidentally true generalizations. Second, there is the matter of how laws connect with the problem of induction. Third is the question of whether laws are sometimes contingent truths or whether they are always necessary truths. (This matter of the modality of laws plays a lesser role than do the other three.) Fourth, there is the topic of whether there really are any strict laws in either fundamental physics or the special sciences. Each of these four issues receives a preliminary presentation below. My hope is that this introduction frames the philosophical questions about laws in a way that helps to make clear the significance of the readings in this book.

1. Lawhood

Here are three reasons philosophers examine what it is to be a law of nature:

1. *Scientific Practice.* Science embraces and employs many principles that were at least once thought to be laws of nature. Newton's law of gravitation, his three laws of motion, and Mendel's laws are a few simple examples.

2. *Related Issues.* The account of counterfactuals championed by Roderick Chisholm (1946, 1955) and by Nelson Goodman (1947) and the deductive-nomological model of explanation advanced by Carl Hempel and Paul Oppenheim (1948) prompted much of the contemporary research on laws. Similarly, in his "The New Riddle of Induction" (1954), Goodman famously claimed that there is a connection between lawhood and induction. Thus, some philosophers following Goodman have an epistemological motivation for studying lawhood.

From *Philosophy and Phenomenological Research* 61 (2000): 571–94.

3. *Philosophers love a good puzzle.* It may have been true that all the cigarettes in A. J. Ayer's cigarette case at some time *t* were made of Virginian tobacco (1956, 144), but this would not thereby have been a law of nature; even if true, this generalization would have been too accidental. Einstein's principle that no signals travel faster than light is also a true generalization but, in contrast, it is thought to be a law. Philosophers rightly wonder: What makes the difference?

What *does* make the difference? The generalization about the cigarettes is a restricted one in that it is about a specific time and a particular person's cigarette case; the principle of relativity is not similarly restricted. So, it is easy to think that, unlike laws, accidentally true generalizations are about specific times or objects. But that is not what makes the difference. Consider the generalization that all gold spheres are less than one mile in diameter. It is unrestricted and there are no mile-in-diameter gold spheres, but this is still not a law. The truly perplexing nature of the puzzle is clearly revealed when the gold-sphere generalization is paired with a similar generalization about uranium spheres:

All gold spheres are less than a mile in diameter.

All uranium spheres are less than a mile in diameter.

Though the former is not a law, the latter arguably is. The latter is not nearly so accidental as the first, since uranium's critical mass is such as to guarantee that enormous spheres of uranium will never exist (cf. van Fraassen 1989, 27). What makes this difference? What makes the former an accidental generalization and the latter a law?

One popular answer ties being a law to deductive systems. The idea dates back to John Stuart Mill (1947, first published 1843), but was defended by Frank Ramsey (1978, first published 1928) and made popular by David Lewis (1973, 1983, 1986, 1994) and John Earman (1984). A version of this idea is defended here in Barry Loewer's "Humean Supervenience" (Selection 10). According to Lewis (1973, 73), the laws of nature belong to all the true deductive systems with a *best* combination of simplicity and strength.[1] So, for example, as Loewer points out, the thought is that the uranium-spheres generalization is a law because it is part of the best deductive systems; quantum theory is an excellent theory of our universe and might be part of the best systems, and it is plausible to think that quantum theory plus truths describing the nature of uranium would entail that there are no mile-in-diameter uranium spheres. The thought is also that the gold-spheres generalization is not a part of the best deductive systems. Of course, that generalization could be added as an axiom to any system, but not without sacrificing simplicity.

Other features of the systems approach are appealing. For one thing, it is reasonable to think that one goal of scientific theorizing is the formulation of true theories that are well balanced in terms of simplicity and strength. So, the systems approach seems to underwrite the truism that an aim of science is the discovery of laws. For another, it is appealing to many that the systems approach is in keeping with broadly Humean constraints on an account of lawhood in that there is no overt appeal to closely related modal concepts (for example, the counterfactual conditional) and no overt appeal to modality-supplying entities (for example, possible worlds or universals). Indeed, the systems account is crucial to Lewis's defense of the principle he calls *Humean supervenience,* "the doctrine that all there is to the world is a vast mosaic of local matters of particular fact, just one little thing and then another" (1986, ix).

In the late 1970s, there emerged competition for the systems approach and all other Humean attempts to say what it is to be a law. Led by David Armstrong (1978, 1983, 1991, 1993), and represented here by Fred Dretske's "Laws of Nature" (Selection 1) and Michael Tooley's "The Nature of Laws" (Selection 2), this rival approach appeals to universals to distinguish laws from nonlaws. Focusing on Armstrong's development of the view, here is one of his concise statements of the framework characteristic of the universals approach:

> Suppose it to be a law that Fs are Gs. F-ness and G-ness are taken to be universals. A certain relation, a relation of nonlogical or contingent necessitation, holds between F-ness and G-ness. This state of affairs may be symbolized as "N(F,G)." (1983, 85)

This framework promises to address our puzzle: Maybe the difference between the uranium-spheres generalization and the gold-spheres generalization is that being uranium does necessitate being less than one mile in diameter, but being gold does not. In sharp contrast to the systems approach, however, the universals approach is not a Humean solution to the problem of laws. Indeed, *non*supervenience is embraced by the universals approach.

The issue of supervenience is so divisive it will be worthwhile to consider two of the arguments that move philosophers to abandon Humeanism. Tooley (Selection 2) ingeniously asks us to suppose that there are ten different kinds of fundamental particles. So, there are fifty-five possible kinds of two-particle interactions. Suppose that fifty-four of these kinds have been studied and fifty-four laws have been discovered. The interaction of X and Y particles have not been studied because conditions are such that they never will interact. Nevertheless, it seems that it might be a law that, when X particles and Y particles interact, P occurs. Similarly, it might be a law that when X and Y particles interact, Q occurs. There seems to be nothing nonnomic about the particulars of this world that fixes which of these generalizations is a law.

Following Tooley's lead, though not pulled to the universals approach, my own nonsupervenience argument (1994, 60–68) begins with a possible world, U_1. Focusing on one specific moment t_0, let's suppose that there is a specific X particle, b, that is subject to a Y field. It has spin up; b's behavior is in no way exceptional. Whenever an X particle enters a Y field, it acquires spin up. Everyone would agree that L_1, the generalization that all X particles subject to a Y field have spin up, could be a law of U_1. U_2 is very similar to U_1. X particles that enter Y fields even do so at exactly the same time and place that they do in U_1. So, for instance, b enters that same Y field at time t_0. What is new about U_2 is that when b enters the Y field at time t_0 it does not acquire spin up. Of course, there must be at least one more difference between these two worlds. Though L_1 could be a law of U_1, L_1 could not be a law of U_2; L_1 is not true in U_2. Now there is nothing particularly remarkable about either U_1 or U_2—nothing to make a Humean suspicious. But here is the catch: It is natural to think that L_1's status as a law in U_1 does not depend on the fact that b entered that Y field at time t_0. That is, even if particle b had not been subject to a Y field at time t_0, L_1 would surely still be a law. Yet it is just as natural to think that L_1's status as a nonlaw in U_2 does not depend on the fact that b entered that Y field at time t_0. L_1 would not be a law in U_2 even if particle b had not been subject to that Y field. Just as you cannot *prevent* L_1 from being a law in U_1 by stopping b from entering that Y field, you cannot *make* L_1 a law by doing the same in U_2. So, it seems that there are two more possible worlds. In one, U_{1*}, X particles are subject to a Y field, all of them have spin up, and L_1 is a law. In the other, U_{2*}, X particles are subject to a Y field, all of them have spin up, but L_1 is not a law. Like Tooley's case, my U_{1*} and U_{2*} constitute an apparent counterexample to supervenience.

Though the universals approach and certain other views can take Tooley's and my examples at face value, Humeans must contend that these so-called possible worlds are not both really possible. Challenges to nonsupervenience arguments from the Humean camp are raised by Helen Beebee in "The Nongoverning Conception of Laws of Nature" (Selection 12) and more briefly by Loewer (Selection 10). Beebee accuses Tooley and I of begging the question in virtue of assuming a *governing* conception of laws.

Two views put forward in the wake of the universals and systems approaches are the no-laws view and antireductionism:

1. *No Laws.* The majority of contemporary philosophers agree that there are laws. There are, however, philosophers who disagree. For example, at the end of his critical discussion of the universals approach in "Armstrong on Laws and Probability" (Selection 6), Bas van Fraassen anticipates the view he develops at length in his *Laws and Symmetries*—the view that there are no laws of nature. Van Fraassen finds support for this in the problems facing accounts of lawhood (for example, Lewis's and Armstrong's), and the perceived failure of philosophers to describe an adequate epistemology that permits rational belief in laws (van Fraassen 1989, 130, 180–81). Ronald Giere also finds support for this view in the origins of the use of the concept of law in the history of science (1995, 122–27), and in his belief that the generalizations often described as laws are not in fact true (127–30).

2. *Antireductionism.* Marc Lange (2000) and I (1994) advocate antireductionist views (also see Woodward 1992). Regarding the question, "What is it to be a law of nature?" we reject the answers given by Humeans and see no advantage in an appeal to universals. We reject all attempts to say what it is to be a law that do not appeal to nomic concepts. Yet we still believe that there are laws; we are not eliminativists à la Giere and van Fraassen. My focus is on arguing for the nonsupervenience of lawhood and highlighting the interconnections between lawhood and other related concepts. Lange's treatment includes an account of what it is to be a law in terms of a counterfactual notion of stability.

There are many, many other attempts to say (or—for principled reasons—not to say) what it is to be a law in the contemporary philosophical literature. Robert Pargetter's (1984) possible worlds account (also see Bigelow and Pargetter 1990, 214–62) and Simon Blackburn's (1984 and 1986) quasi-realism (also see Ward 2002) are just two more important examples.

That, anyway, is how the lines have been drawn about lawhood. What will be interesting is how matters will progress. How will philosophy move beyond the current divisions in logical space? Make no mistake, the divisions are serious ones: supervenience versus nonsupervenience, sparse ontology versus lush ontology, laws versus no laws, reducibility versus irreducibility. These are all major issues about which leading philosophers have made quite contrary judgments both about individual examples and about methodology. Do laws govern or just describe? Does scientific practice or common sense judgments provide the ultimate testing ground? How primary a role should epistemology or semantics play in the formulation of the correct metaphysics of lawhood? New work will have to address these debates head-on. The subsequent collision should generate new fruitful territory.

2. Induction

Goodman thought that the difference between laws of nature and accidental truths was linked inextricably with the problem of induction. In "The New Riddle of Induction," Goodman says,

> Only a statement that is *lawlike*—regardless of its truth or falsity or its scientific importance—is capable of receiving confirmation from an instance of it; accidental statements are not (1954, 74).

(Terminology: P is lawlike if and only if P is a law if true.) Goodman claims that, if a generalization is accidentally true (and so not lawlike), then it is not capable of receiving confirmation from one of its instances.

This has prompted much discussion, including some challenges. For example, Dretske (Selection 1) asks us to consider ten flips of a fair coin, and to suppose that the first nine land heads. The first nine instances—at least in a sense—confirm the generalization that all the flips will land heads; the probability of that generalization is raised from $(.5)^{10}$ up to .5. But this generalization is not lawlike; if true, it is not a law. So accidental statements are capable of receiving confirmation. It is standard to respond to such a challenge by arguing that this is not the pertinent sense of confirmation (that it is mere "content-cutting"), and by suggesting that what does require lawlikeness is confirmation of the generalization's unexamined instances. Notice that, in Dretske's coin case, the probability that the tenth flip will land heads does not change after the first nine flips. There are, however, other examples that generate a challenge for this idea, too. In "Confirmation and the Nomological" (Selection 4), Frank Jackson and Robert Pargetter ask us to consider a room with one hundred men. By directly questioning the first fifty you encounter, you find out that each one of the fifty is a third son. Intuitively, it seems that this would increase your reason to think that all the men in this room are third sons *and* that the next person you encounter will also be a third son. It does no good to revise Goodman's claim to say that no generalization *believed* to be accidental is capable of confirmation. About the third-son case, one would know that the generalization that all the men in this room are third sons, even if true, would not be a law. The discussion continues. Jackson and Pargetter have proposed an alternative connection between confirmation and laws on which certain counterfactual truths must hold. This suggestion is criticized by Elliott Sober in his "Confirmation and Law-likeness" (Selection 7).[2]

Sometimes the idea that laws have a special role to play in induction serves as starting point for a criticism of Humean analyses. Dretske (Selection 1) and Armstrong (1983, 52–59; 1991) adopt a model of inductive inference on which it involves an inference to the best explanation. This model is defended by John Foster in "Induction, Explanation, and Natural Necessity" (Selection 5). On its simplest construal, the model describes a pattern that begins with an observation of instances of a generalization, includes an inference to there being the corresponding law (this is the inference to the best explanation), and concludes with an inference to the generalization itself. The complaint lodged against Humeans is that, on their view of what laws are, laws are not suited to explain their instances and so cannot sustain the required inference to the best explanation. As Dretske and Armstrong see it, on their own views, laws are suited to play the required explanatory role, since on their views a law is not just a universal generalization, but is an entirely different creature—a relation holding between two universals.

Armstrong and Dretske make substantive claims on what can and cannot be instance confirmed: roughly, Humean laws cannot, laws-as-universals can. But, at the very least, these claims cannot be quite right. Humean laws cannot? As the discussion above illustrates, even generalizations known to be accidental can be confirmed by their instances. Dretske and Armstrong need a plausible and suitably strong premise connecting lawhood to confirmability and it is not clear that there is one to be had. Here is the basic problem: As Sober and others have pointed out, the confirmation of a hypothesis or its unexamined instances will always be sensitive to what background beliefs are in place. So much so, with background beliefs of the right sort, just about anything can be confirmed irrespective of its status as a law or whether it is lawlike. Thus, stating a plausible principle describing the connection between laws of nature and the problem of induction will be difficult. In order to uncover a nomological constraint on induction, something needs to be said about the role of background beliefs.

3. Necessity

Philosophers have generally held that some contingent truths are laws of nature. Furthermore, they have thought that if it is a law that all Fs are Gs, then there need not be any (metaphysically) necessary connection between F-ness and G-ness, that it is possible that something be F without being G. For example, any possible world that as a matter of law obeys the general principles of Newtonian physics is a world in which Newton's first is true and a world containing accelerating inertial bodies is a world in which Newton's first is false. The latter world is also a world where being inertial is instantiated but does not necessitate no acceleration. Some *necessitarians,* however, hold that all laws of nature are necessary truths (see Shoemaker 1980, 1998; Swoyer 1982; and Fales 1990). Other necessitarians have held something that is only slightly different. Maintaining that some laws are singular statements about universals, they allow that some laws are contingently true. So, on this view, an F-ness/G-ness law could be false if F-ness does not exist. Still, this difference really is minor. These authors think that, for there to be an F-ness/G-ness law, it must be necessarily true that all Fs are Gs. This view is advanced here in John Bigelow, Brian Ellis, and Caroline Lierse's "The World as One of a Kind: Natural Necessity and Laws of Nature" (Selection 8). In support of their view, necessitarians often argue that their position is a consequence of the correct theory of property individuation. Roughly, mass just would not be the property it is unless it had the causal powers it does, and hence obeyed the laws that it does. As they see it, it is also a virtue of their position that they can account for why laws are counterfactual supporting: They support counterfactuals in the same way that the truths of logic and mathematics do (Swoyer 1982, 209; Fales 1990, 85–87).[3]

4. Physics and the Special Sciences

Two separate (but related) questions that are not always clearly distinguished have received much recent attention in the philosophical literature surrounding laws of nature. Neither has much directly to do with what it is to be a law. Instead, they have to do with the nature of the generalizations scientists try to discover. First: Does *any* science try to discover exceptionless regularities in its attempt to discover laws of nature? Second: Even if one science—fundamental physics—does, do others?

Philosophers draw a distinction between what they call *strict generalizations* and *ceteris-paribus generalizations*. The contrast is supposed to be between universal generalizations of the sort discussed above (for example, that no signals travel faster than light) and seemingly less strict generalizations such as: other things being equal, all raptors have a hooked beak. The idea is that the former would be contradicted by a single counterinstance, say, one superluminal signal, though the latter is consistent with there being a deformed or injured raptor without any beak at all. Though in theory this distinction is easy enough to understand, in practice it is often difficult to distinguish strict from ceteris-paribus generalizations. This is because many philosophers think that many utterances which include no explicit ceteris-paribus clause implicitly do include such a clause.

For the most part, philosophers have thought that if scientists have discovered any exceptionless regularities that are laws of nature, they have done so at the level of fundamental physics. A few philosophers, however, are doubtful that there are exceptionless regularities at even this basic level. For example, in "Do the Laws of Physics State the Facts?" (Selection 3), Nancy Cartwright argues that the descriptive and the explanatory aspects of laws conflict. Consider Newton's gravitational principle, $F = Gmm'/r^2$. Properly understood, according to Cartwright, it says that for any two bodies the force between them is Gmm'/r^2. But if that is what the law says then the law is not an exceptionless regularity. This is because the force between two bodies is influenced by things other than just their mass and the distance between them (such as the charge of the two bodies as described by Coulomb's law). The statement of the gravitational principle can be amended to make it true, but that, according to Cartwright, at least on certain standard ways of doing so, would strip it of its explanatory power. For example, if the principle is taken to hold only that $F = Gmm'/r^2$ if there are no forces other than gravitational forces at work, then it would be true but would not apply except in idealized circumstances. In his "Natural Laws and the Problem of Provisos" (Selection 9), Marc Lange makes a similar point. Consider a standard expression of the law of thermal expansion: "Whenever the temperature of a metal bar of length L_0 changes by ΔT, the length of the bar changes by $\Delta L = kL_0 \Delta T$," where k is a constant, the thermal expansion coefficient of the metal. If this expression were used to express the strict generalization straightforwardly suggested by its grammar, then such an utterance would be false since the length of a bar does not change in the way described in cases where someone is hammering on its ends. It looks like the law will require provisos, but so many that the only apparent way of taking into consideration all the required provisos would be with something like a ceteris-paribus clause. Then the concern becomes that the statement would be empty. Because of the difficulty of stating plausible truth conditions for ceteris-paribus sentences, it is feared that, "Other things being equal, $\Delta L = kL_0 \Delta T$" could only mean "$\Delta L = kL_0 \Delta T$ provided that $\Delta L = kL_0 \Delta T$."

Even those who agree with the arguments of Lange and Cartwright sometimes disagree about what ultimately the arguments say about laws. Cartwright believes that the true laws of nature are not exceptionless regularities, but instead are statements that describe causal powers. So construed, they turn out to be both true and explanatory. Lange ends up holding that there are propositions properly adopted as laws though in doing so one need not also believe any exceptionless regularity; there need not be one. Ronald Giere (1995) can usefully and accurately be thought of as agreeing with Cartwright's and Lange's basic arguments but insisting that law statements do not have implicit provisos or implicit ceteris-paribus clauses. So, he concludes that there are no laws.

In "*Ceteris Paribus,* There Is No Problem of Provisos" (Selection 11), John Earman and John Roberts hold that there are exceptionless and lawful regularities. More precisely, they argue that scientists involved in fundamental physics do attempt to state strict generalizations and are such that these generalizations would be laws if they were true:

> Our claim is only that . . . typical theories from fundamental physics are such that *if* they were true, there would be precise proviso free laws. For example, Einstein's gravitational field law asserts—without equivocation, qualification, proviso, *ceteris paribus* clause—that the Ricci curvature tensor

> of spacetime is proportional to the total stress-energy tensor for matter-energy; the relativistic version of Maxwell's laws of electromagnetism for charge-free flat spacetime asserts—without qualification or proviso—that the curl of the **E** field is proportional to the partial time derivative, etc.

About Cartwright's example, they say that a plausible understanding of the gravitational principle is as describing only the *gravitational* force between the two massive bodies. (Cartwright argues that there is no such component force and so thinks such an interpretation would be false. Earman and Roberts disagree.) About Lange's example, they say the law should be understood as having the single proviso that there be no external stresses on the metal bar. In any case, much more would need to be said to establish that *all* the apparently strict and explanatory generalizations that have been or will be stated in physics have turned or will turn out to be false.[4]

Supposing that physicists do try to discover exceptionless regularities, and even supposing that our physicists will sometimes be successful, there is a further question of whether it is a goal of any science other than fundamental physics—any so-called special science—to discover exceptionless regularities and whether these scientists have any hope of succeeding.[5] Consider an economic law of supply and demand that says that, when demand increases and supply is held fixed, price increases. Notice that, in some places, the price of gasoline has sometimes remained the same despite an increase of demand and a fixed supply, because the price of gasoline was government regulated. It appears that the law has to be understood as having a ceteris-paribus clause in order for it to be true. This problem appears to be a very general one. As Jerry Fodor (1989, 78) has pointed out, in virtue of being stated in a vocabulary of a special science, it is very likely that there will be limiting conditions—especially underlying physical conditions—that will undermine any interesting strict generalization of the special sciences, conditions that themselves could not be described in the special-science vocabulary.

As I see it, progress on the problem of provisos requires three basic issues to be distinguished. First, there is the question of what it is to be a law, which in essence is the search for a necessarily true completion of: "P is a law if and only if. . . ." Obviously, to be a true completion, it must hold for all P, whether P is a strict generalization or a ceteris-paribus one. Second, there is also a need to determine the truth conditions of the generalization sentences used by scientists. Third, there is the a posteriori and scientific question of which generalizations expressed by the sentences used by the scientists are true. The second of these issues is where the action needs to be.

On this score, it is striking how little attention is given to the possible effects of context. Might it not be that, when the economist utters a certain strict generalization sentence in an "economic setting" (say, in an economics textbook or at an economics conference), context-sensitive considerations affecting its truth conditions will have it turn out that the utterance is true? This might be the case despite the fact that the same sentence uttered in a different context (say, in a discussion among fundamental physicists or better yet in a philosophical discussion of laws) would result in a clearly false utterance. These changing truth conditions might be the result of something as plain as a contextual shift in the domain of quantification or perhaps something less obvious. Whatever it is, the important point is that this shift could be a function of nothing more than the linguistic meaning of the sentence and familiar rules of interpretation.

Consider a situation where an engineering professor utters, "When a metal bar is heated the change in its length is proportional to the change in its temperature," and suppose a student offers, "Not when someone is hammering on both ends of the bar." Has the student shown that the teacher's utterance was false? Maybe not. Notice that the student comes off sounding at least a little bit insolent. In all likelihood, such an unusual situation as someone hammering on both ends of a heated bar would not have been in play when the professor said what he did. In fact, the reason the student comes off sounding somewhat insolent is because it seems that he should have known that his example was irrelevant. Notice that the professor's sentence need not include some implicit ceteris-paribus clause in order for his utterance to be true; in ordinary conversation, plain old strict generalization sentences are rarely used to cover the full range of actual cases.[6]

If special scientists do make true utterances of generalization sentences (sometimes ceteris-paribus generalizations sentences, sometimes not), apparently nothing stands in the way of them uttering true special-science lawhood sentences. The issue here has been the truth of special-science generalizations, not any other requirements of lawhood. Just like the generalization sentences, these lawhood sentences will be false in plenty of contexts, but that is inconsequential.

5. On What Is to Come

The readings contained in the rest of this collection are meant to reflect the contemporary history of the problem of laws, but are not intended to be a thorough survey of the literature on laws, not even the most recent (post-1976) literature. No doubt the choice of selections was dictated by some of my own views about what are the most interesting and important issues. My judgment was that this was the best way to put together a coherent package of readings. The essays have been ordered chronologically based on their first date of publication, and my hope is that this format will give the reader a feel for the recent historical development of the problem.[7]

Notes

1. Lewis later (1986, 1994) made significant revisions to his account in order to address problems involving physical probability.

2. Marc Lange (2000, 111–42) uses a different strategy. He tries to refine further the relevant notion of confirmation, characterizing what he takes to be an intuitive notion of inductive confirmation, and then contends that only generalizations that are not believed not to be lawlike can be (in his sense) inductively confirmed.

3. Alan Sidelle challenges this argument in his "On the Metaphysical Contingency of Laws of Nature" (2002).

4. Volume 57 (2002) of *Erkenntnis* is devoted to the topic of ceteris-paribus laws. It includes new papers by Cartwright and Lange.

5. Donald Davidson prompted much of the recent interest in special-science laws in his "Mental Events" (1980, first published 1970). He gave an argument specifically directed against the possibility of strict psycho-physical laws. More importantly, he made the suggestion that the absence of such laws may be relevant to whether mental events ever cause physical events. This prompted a slew of papers dealing with the problem of reconciling the absence of strict special-science laws with the reality of mental causation (see Loewer and Lepore 1987, 1989; Fodor 1989, 1991; Schiffer 1991; Pietroski and Rey 1995).

6. Ceteris-paribus clauses may be overused in the special sciences and elsewhere. Sometimes speakers fail to appreciate the context sensitivity of the sentences they use. We all sometimes feel the need to hedge what we say in order to head off the nitpickers. My suggestion is that *sometimes* the nitpickers are out of line and are nothing more than context changers.

7. The introduction is based on the original version of my entry on laws of nature for *The Stanford Encyclopedia of Philosophy*. This entry will be regularly updated and so will provide a useful way to stay current with the literature.

References

Armstrong, D. 1978. *A Theory of Universals*. Cambridge: Cambridge University Press.
——. 1983. *What Is a Law of Nature?* Cambridge: Cambridge University Press.
——. 1991. "What Makes Induction Rational?" *Dialogue* 30: 503–11.
——. 1993. "The Identification Problem and the Inference Problem." *Philosophy and Phenomenological Research* 53: 421–22.
Ayer, A. 1956. "What Is a Law of Nature?" *Revue Internationale de Philosophia* 10: 144–65.
Bigelow, J., and Pargetter, R. 1990. *Science and Necessity*. Cambridge: Cambridge University Press.
Blackburn, S. 1984. *Spreading the Word*. Oxford: Clarendon Press.
——. 1986. "Morals and Modals." In *Fact, Science and Morality,* edited by G. Macdonald and C. Wright. Oxford: Basil Blackwell.
Carroll, J. 1994. *Laws of Nature*. Cambridge: Cambridge University Press.
——. 2003. "Laws of Nature." *The Stanford Encyclopedia of Philosophy* (Summer 2003 edition), edited by E. Zalta, http://plato.stanford.edu/archives/sum2003/entries/laws-of-nature/.
Chisholm, R. 1946. "The Contrary-to-Fact Conditional." *Mind* 55: 289–307.
——. 1955. "Law Statements and Counterfactual Inference." *Analysis* 15: 97–105.
Davidson, D. 1980. *Essays on Actions and Events*. Oxford: Clarendon Press.
Earman, J. 1984. "Laws of Nature: The Empiricist Challenge." In *D. M. Armstrong,* edited by R. Bogdan. Dordrecht: D. Reidel Publishing Company.
Fales, E. 1990. *Causation and Universals*. London: Routledge.
Fodor, J. 1989. "Making Mind Matter More." *Philosophical Topics* 17: 59–79.
——. 1991. "You Can Fool Some of the People All the Time, Everything Else Being Equal: Hedged Laws and Psychological Explanations." *Mind* 100: 19–34.
Giere, R. 1995. "The Skeptical Perspective: Science without Laws of Nature." In *Laws of Nature,* edited by F. Weinert. Berlin: Walter de Gruyter.
Goodman, N. 1947. "The Problem of Counterfactual Conditionals." *Journal of Philosophy* 44: 113–28.
——. 1954. *Fact, Fiction, and Forecast*. London: The Athlone Press.
Kneale, W. 1961. "Universality and Necessity." *British Journal for the Philosophy of Science* 12: 89–102.
Lange, M. 2000. *Natural Laws in Scientific Practice*. Oxford: Oxford University Press.
Lewis, D. 1973. *Counterfactuals*. Cambridge: Harvard University Press.
——. 1983. "New Work for a Theory of Universals." *Australasian Journal of Philosophy* 61: 343–77.
——. 1986. *Philosophical Papers*. Vol. 2. New York: Oxford University Press.
——. 1994. "Chance and Credence: Humean Supervenience Debugged." *Mind* 103: 473–90.
Loewer, B., and Lepore, E. 1987. "Mind Matters." *Journal of Philosophy* 84: 630–42.
——. 1989. "More on Making Mind Matter." *Philosophical Topics* 17: 175–91.
Mill, J. 1947. *A System of Logic*. London: Longmans, Green and Co.
Pargetter, R. 1984. "Laws and Modal Realism." *Philosophical Studies* 46: 335–47.

Pietroski, P., and Rey, G. 1995. "When Other Things Aren't Equal: Saving *Ceteris Paribus* Laws from Vacuity." *British Journal for the Philosophy of Science* 46: 81–110.

Ramsey, F. 1978. *Foundations*. London: Routledge and Kegan Paul.

Schiffer, S. 1991. "Ceteris Paribus Laws." *Mind* 100: 1–17.

Shoemaker, S. 1980. "Causality and Properties." In *Time and Cause,* edited by P. van Inwagen. Dordrecht: D. Reidel Publishing Company.

——. 1998. "Causal and Metaphysical Necessity." *Pacific Philosophical Quarterly* 79: 59–77.

Sidelle, A. 2002. "On the Metaphysical Contingency of Laws of Nature." In *Conceivability and Possibility,* edited by T. Szabó Gendler and J. Hawthorne. Oxford: Clarendon Press.

Swoyer, C. 1982. "The Nature of Natural Laws." *Australasian Journal of Philosophy* 60: 203–23.

Van Fraassen, B. 1989. *Laws and Symmetry*. Oxford: Clarendon Press.

Ward, B. 2002. "Humeanism without Humean Supervenience: A Projectivist Account of Laws and Possibilities." *Philosophical Studies* 107: 191–218.

Woodward, J. 1992. "Realism about Laws." *Erkenntnis* 36: 181–218.

1

Laws of Nature

FRED I. DRETSKE

It is tempting to identify the laws of nature with a certain class of universal truths. Very few empiricists have succeeded in resisting this temptation. The popular way of succumbing is to equate the fundamental laws of nature with what is asserted by those universally true statements of nonlimited scope that embody only qualitative predicates.[1] On this view of things a law-like statement is a statement of the form "$(x)(Fx \supset Gx)$" or "$(x)(Fx \equiv Gx)$" where "F" and "G" are purely qualitative (nonpositional). Those law-like statements that are true express laws. "All robins' eggs are greenish blue," "All metals conduct electricity," and "At constant pressure any gas expands with increasing temperature" (Hempel's examples) are law-like statements. If they are true, they express laws. The more familiar sorts of things that we are accustomed to calling laws, the formulae and equations appearing in our physics and chemistry books, can supposedly be understood in the same way by using functors in place of the propositional functions "Fx" and "Gx" in the symbolic expressions given above.

I say that it is tempting to proceed in this way since, to put it bluntly, conceiving of a law as having a content greater than that expressed by a statement of the form $(x)(Fx \supset Gx)$ seems to put it beyond our epistemological grasp.[2] We must work with what we are given, and what we are given (the observational and experimental data) are facts of the form: this *F* is *G*, that *F* is *G*, all examined *F*'s have been *G*, and so on. If, as some philosophers have argued,[3] law-like statements express a kind of nomic necessity between events, something *more* than that *F*'s are, as a matter of fact, always and everywhere, *G*, then it is hard to see what kind of evidence might be brought in support of them. The whole point in acquiring instantial evidence (evidence of the form "This *F* is *G*") in support of a law-like hypoth-

This essay is reprinted from *Philosophy of Science* 44 (1977): 248–68.

esis would be lost if we supposed that what the hypothesis was actually asserting was some kind of nomic connection, some kind of modal relationship, between things that were *F* and things that were *G*. We would, it seems, be in the position of someone trying to confirm the *analyticity* of "All bachelors are unmarried" by collecting evidence about the marital status of various bachelors. This kind of evidence, though relevant to the *truth* of the claim that all bachelors are unmarried, is powerless to confirm the *modality* in question. Similarly, if a hypothesis, in order to qualify as a law, must express or assert some form of necessity between *F*'s and *G*'s, then it becomes a mystery how we ever manage to confirm such attributions with the sort of instantial evidence available from observation.

Despite this argument, the fact remains that laws are *not* simply what universally true statements express, not even universally true statements that embody purely qualitative predicates (and are, as a result, unlimited in scope). This is not particularly newsworthy. It is commonly acknowledged that law-like statements have some peculiarities that prevent their straightforward assimilation to universal truths. That the concept of a law and the concept of a universal truth are different concepts can best be seen, I think, by the following consideration: assume that $(x)(Fx \supset Gx)$ is true and that the predicate expressions satisfy all the restrictions that one might wish to impose in order to convert this universal statement into a statement of law.[4] Consider a predicate expression *"K"* (eternally) coextensive with *"F"*; i.e., $(x)(Fx \equiv Kx)$ for all time. We may then infer that if $(x)(Fx \supset Gx)$ is a universal truth, so is $(x)(Kx \supset Gx)$. The class of universal truths is closed under the operation of coextensive predicate substitution. Such is *not* the case with laws. If it is a law that all *F*'s are *G*, and we substitute the term *"K"* for the term *"F"* in this law, the result is not necessarily a law. If diamonds have a refractive index of 2.419 (law) and "is a diamond" is coextensive with "is mined in kimberlite (a dark basic rock)" we cannot infer that *it is a law* that things mined in kimberlite have a refractive index of 2.419. Whether this is a law or not depends on whether the coextensiveness of "is a diamond" and "is mined in kimberlite" is *itself* law-like. The class of laws is not closed under the same operation as is the class of universal truths.

Using familiar terminology we may say that the predicate positions in a statement of law are *opaque* while the predicate positions in a universal truth of the form $(x)(Fx \supset Gx)$ are *transparent*. I am using these terms in a slightly unorthodox way. It is not that when we have a law, "All *F*'s are *G*," we can alter its truth value by substituting a coextensive predicate for *"F"* or

"G." For if the statement is true, it will remain true after substitution. What happens, rather, is that the expression's status *as a law* is (or may be) affected by such an exchange. The matter can be put this way: the statement

(A) All *F*'s are *G* (understood as $(x)(Fx \supset Gx)$)

has *"F"* and *"G"* occurring in transparent positions. Its truth value is unaffected by the replacement of *"F"* or *"G"* by a coextensive predicate. The same is true of

(B) It is universally true that *F*'s are *G*.

If, however, we look at

(C) It is a law that *F*'s are *G*.

we find that *"F"* and *"G"* occur in opaque positions. If we think of the two prefixes in (B) and (C), "it is universally true that . . ." and "it is a law that . . . ," as operators, we can say that the operator in (B) does not, while the operator in (C) does, confer opacity on the embedded predicate positions. To refer to something as a statement of law is to refer to it as an expression in which the descriptive terms occupy opaque positions. To refer to something as a universal truth is to refer to it as an expression in which the descriptive terms occupy transparent positions. Hence, our concept of a law differs from our concept of a universal truth.[5]

Confronted by a difference of this sort, many philosophers have argued that the distinction between a natural law and a universal truth was not, fundamentally, an *intrinsic* difference. Rather, the difference was a difference in the *role* some universal statements played within the larger theoretical enterprise. Some universal statements are more highly integrated into the constellation of accepted scientific principles, they play a more significant role in the explanation and prediction of experimental results, they are better confirmed, have survived more tests, and make a more substantial contribution to the regulation of experimental inquiry. But, divorced from this context, stripped of these *extrinsic* features, a law is nothing but a universal truth. It has the same empirical content. Laws are to universal truths what shims are to slivers of wood and metal; the latter *become* the former by being *used* in a certain way. There is a *functional* difference, nothing else.[6]

According to this reductionistic view, the peculiar opacity (described above) associated with laws is not a manifestation of some intrinsic difference between a law and a universal truth. It is merely a symptom of the special status or function that some universal statements have. The basic formula is: law = universal truth + *X*. The *"X"* is intended to indicate the special function, status, or role that a universal truth must have to qualify as a law. Some popular candidates for this auxiliary idea, *X,* are:

(1) High degree of confirmation,
(2) Wide acceptance (well established in the relevant community),
(3) Explanatory potential (can be used to explain its instances),
(4) Deductive integration (within a larger system of statements),
(5) Predictive use.

To illustrate the way these values of *X* are used to buttress the equation of laws with universal truths, it should be noted that each of the concepts appearing on this list generates an opacity similar to that witnessed in the case of genuine laws. For example, to say that it is a law that all *F*'s are *G* may possibly be no more than to say that it is well established that $(x)(Fx \supset Gx)$. The peculiar opacity of laws is then explained by pointing out that the class of expressions that are well established (or highly confirmed) is not closed under substitution of coextensive predicates: one cannot infer that $(x)(Kx \supset Gx)$ is well established just because *"Fx"* and *"Kx"* are coextensive and $(x)(Fx \supset Gx)$ is well established (for no one may know that *"Fx"* and *"Kx" are* coextensive). It may be supposed, therefore, that the opacity of laws is merely a manifestation of the underlying fact that a universal statement, to qualify as a law, must be well established, and the opacity is a result of this epistemic condition. Or, if this will not do, we can suppose that one of the other notions mentioned above, or a combination of them, is the source of a law's opacity.

This response to the alleged uniqueness of natural laws is more or less standard fare among empiricists in the Humean tradition. Longstanding (= venerable) epistemological and ontological commitments motivate the equation: law = universal truth + *X*. There is disagreement among authors about the differentia *X,* but there is near unanimity about the fact that laws are a *species* of universal truth.

If we set aside our scruples for the moment, however, there is a plausible explanation for the opacity of laws that has not yet been mentioned. Taking our cue from Frege, it may be argued that since the operator "it is a law that . . ." converts the otherwise transparent positions of "All *F*'s are *G*" into opaque positions, we may conclude that this occurs because within the context of this operator (either explicitly present or implicitly understood) the terms *"F"* and *"G"* do not have their usual referents. There is a shift in what we are talking about. To say that *it is a law* that *F*'s are *G* is to say that "All *F*'s are *G*" is to be understood (insofar as it expresses a law), not as a statement about the extensions of the predicates *"F"* and *"G,"* but as a singular statement describing a relationship between the universal properties *F*-ness and *G*-ness. In other words, (C) is to be understood as having the form:

(6) *F*-ness → *G*-ness.[7]

To conceive of (A) as a universal truth is to conceive of it as expressing a relationship between the extensions of its terms; to conceive of it as a law is to conceive of it as expressing a relationship between the properties (magnitudes, quantities, features) which these predicates express (and to which we may refer with the corresponding abstract singular term). The opacity of laws is merely a manifestation of this change in reference. If *"F"* and *"K"* are coextensive, we cannot substitute the one for the other in the *law* "All *F*'s are *G*" and expect to preserve truth; for the law asserts a connection between *F*-ness and *G*-ness and there is no guarantee that a similar connection exists between the properties *K*-ness and *G*-ness just because all *F*'s are *K* and *vice versa.*[8]

It is this view that I mean to defend in the remainder of this essay. Law-like statements are singular statements of fact describing a relationship between properties or magnitudes. Laws are the relationships that are asserted to exist by true law-like statements. According to this view, then, there is an *intrinsic* difference between laws and universal truths. Laws imply universal truths, but universal truths do not imply laws. Laws are (expressed by) *singular* statements describing the relationships that exist between universal qualities and quantities; they are not universal statements about the particular objects and situations that exemplify these qualities and quantities. Universal truths are not transformed into laws by acquiring some of the extrinsic properties of laws, by being used in explanation or prediction, by being made to support counterfactuals, or by becoming well established. For, as we shall see, universal truths *cannot* function in these ways. They *cannot* be made to perform a service they are wholly unequipped to provide.

In order to develop this thesis it will be necessary to overcome some metaphysical prejudices, and to overcome these prejudices it will prove useful to review the major deficiencies of the proposed alternative. The attractiveness of the formula: law = universal truth + *X*, lies, partly at least, in its ontological austerity, in its tidy portrayal of what there is, or what there must be, in order for there to be laws of nature. The antidote to this seductive doctrine is a clear realization of how utterly hopeless, epistemologically and functionally hopeless, this equation is.

If the auxiliary ideas mentioned above (explanation, prediction, confirmation, etc.) are deployed as values of *X* in the reductionistic equation of laws with universal truths, one can, as we have already seen, render a satisfactory account of the opacity of laws. In this particular respect the attempted equation proves adequate. In what way, then, does it fail?

(1) and (2) are what I will call "epistemic" notions; they assign to a statement a certain epistemological status or cognitive value. They are, for this reason alone, useless in understanding the nature of a law.[9] Laws do not begin to be laws only when we first become aware of them, when the relevant hypotheses become well established, when there is public endorsement by the relevant scientific community. The laws of nature are the same today as they were one thousand years ago (or so we believe); yet, some hypotheses are highly confirmed today that were not highly confirmed one thousand years ago. It is certainly true that we only begin to *call* something a law when it becomes well established, that we only recognize something as a statement of law when it is confirmed to a certain degree, but that something *is* a law, that some statement does in fact express a law, does not similarly await our appreciation of this fact. We discover laws, we do not invent them—although, of course, some invention may be involved in our manner of expressing or codifying these laws. Hence, the status of something as a statement of law does not depend on its epistemological status. What does depend on such epistemological factors is our ability to identify an otherwise qualified statement *as true* and, therefore, *as a statement of law.* It is for this reason that one cannot appeal to the epistemic operators to clarify the nature of laws; they merely confuse an epistemological with an ontological issue.

What sometimes helps to obscure this point is the tendency to conflate laws with the verbal or symbolic expression of these laws (what I have been calling "statements of law"). Clearly, though, these are different things and should not be confused. There are doubtless laws that have not yet (or will never) receive symbolic expression, and the same law may be given different verbal codifications (think of the variety of ways of expressing the laws of thermodynamics). To use the language of "propositions" for a moment, a law is the proposition expressed, not the vehicle we use to express it. The *use* of a sentence *as an expression of law* depends on epistemological considerations, but the law itself does not.

There is, furthermore, the fact that whatever auxiliary idea we select for understanding laws (as candidates for *X* in the equation: law = universal truth + *X*), if it is going to achieve what we expect of it, should help to account for the variety of other features that laws are acknowledged to have. For example, it is said that laws "support" counterfactuals of a certain sort. If laws are universal truths, this fact is a complete mystery, a mystery that is usually suppressed by using the word "support." For, of course, universal statements do not *imply* counterfactuals in any sense of the word "imply" with which I am familiar. To be told that all *F*'s are *G* is not to be told anything that implies that if this *x* were an *F*, it would be *G*. To be told that all dogs born at sea have been and will be cocker spaniels is *not* to be told that we would get cocker spaniel pups (or no pups at all) if we arranged to breed dachshunds at sea. The only reason we might *think* we were being told this is because we do not expect anyone to assert that all dogs born at sea *will be* cocker spaniels unless they know (or have good reasons for believing) that this is true; and we do not understand *how* anyone could *know* that this is true without being privy to information that insures this result—without, that is, knowing of some bizarre law or circumstance that *prevents* anything but cocker spaniels from being born at sea. Hence, if we accept the claim at all, we do so with a certain presumption about what our informant must know in order to be a serious claimant. We assume that our informant knows of certain laws or conditions that *insure* the continuance of a past regularity, and it is this presumed knowledge that we exploit in endorsing or accepting the counterfactual. But the simple fact remains that the statement "All dogs born at sea have been and will be cocker spaniels" does not *itself* support or imply this counterfactual; at best, *we* support the counterfactual (if we support it at all) on the basis of what the claimant is supposed to know in order to advance such a universal projection.

Given this incapacity on the part of universal truths to support counterfactuals, one would expect some assistance from the epistemic condition if laws are to be analyzed as well-established universal truths. But the expectation is disappointed; we are *left* with a complete mystery. For if a statement of the form "All *F*'s are *G*" does not support the counterfactual, "If this (non-*G*) were an *F*, it would be *G*," it is clear that it will not support it just because it is well established or highly confirmed. The fact that all the marbles in the bag are red does not support the contention that if this (blue) marble were in the bag, it would be red; but neither does the fact that we *know* (or it is highly confirmed) that all the marbles in the bag are red support the claim that if this marble were in the bag it would be red. And making the universal truth *more universal* is not going to repair the difficulty. The fact that all the marbles in the universe are (have been and will be) red does not imply that I *cannot* manufacture a blue marble; it implies that I *will not,* not that I cannot or that if I were to try, I would fail. To represent laws on the model of one of our epistemic operators, therefore, leaves wholly unexplained one of the most important features of laws that we are trying to understand. They are, in this respect, unsatisfactory candidates for the job.

Though laws are not merely well-established general truths, there is a related point that deserves mention: laws are the *sort* of thing that can become well established prior to an exhaustive enumeration of the instances to which they apply. This, of course, is what gives laws their predictive utility. Our confidence in them increases at a much more rapid rate than does the ratio of favorable examined cases to total number of cases. Hence, we reach the point of confidently using them to project the outcome of unexamined situations while there is still a substantial number of unexamined situations to project.

This feature of laws raises new problems for the reductionistic equation. For, contrary to the argument in the second paragraph of this essay, it is hard to see how confirmation is possible for universal truths. To illustrate this difficulty, consider the (presumably easier) case of a general truth of *finite* scope. I have a coin that you have (by examination and test) convinced yourself is quite normal. I propose to flip it ten times. I conjecture (for whatever reason) that it will land heads all ten times. You express doubts. I proceed to "confirm" my hypothesis. I flip the coin once. It lands heads. Is this evidence that my hypothesis is correct? I continue flipping the coin and it turns up with nine straight heads. Given the opening assumption that we are dealing with a fair coin, the probability of getting all ten heads (the probability that my hypothesis is true) is now, after examination of 90% of the total population to which the hypothesis applies, exactly .5. If we are guided by probability considerations alone, the likelihood of all ten tosses being heads is now, after nine favorable trials, a toss-up. After nine favorable trials it is no more reasonable to believe the hypothesis than its denial. In what sense, then, can we be said to have been accumulating evidence (during the first nine trials) that all would be heads? In what sense have we been confirming the hypothesis? It would appear that the probability of my conjecture's being true never exceeds .5 until we have exhaustively examined the entire population of coin tosses and found them *all* favorable. The probability of my conjecture's being true is either: (i) too low (≤ .5) to invest any confidence in the hypothesis, or (ii) so high (= 1) that the hypothesis is useless for prediction. There does not seem to be any middle ground.

Our attempts to confirm universal generalizations of nonlimited scope is, I submit, in exactly the same impossible situation. It is true, of course, that after nine successful trials the probability that all ten tosses will be heads is greatly increased over the initial probability that all would be heads. The initial probability (assuming a fair coin) that all ten tosses would be heads was on the order of .002. After nine favorable trials it is .5. In this sense I have increased the probability that my hypothesis is true; I have raised its probability from .002 to .5. The important point to notice, however, is that this sequence of trials did not alter the probability that the *tenth* trial would be heads. The probability that the unexamined instance would be favorable remains exactly what it was before I began flipping the coin. It was originally .5 and it is now, after nine favorable trials, still .5. I am in no better position now, after extensive sampling, to predict the outcome of the tenth toss than I was before I started. To suppose otherwise is to commit the converse of the Gambler's Fallacy.

Notice, we could take the first nine trials as evidence that the tenth trial would be heads *if* we took the results of the first nine tosses as evidence that the coin was biased in some way. Then, on *this* hypothesis, the probability of getting heads on the last trial (and, hence, on all ten trials) would be greater than .5 (how much greater would depend on the conjectured degree of bias and this, in turn, would presumably depend on the extent of sampling). This new hypothesis, however, is something quite different than the original one. The original hypothesis was of the form: $(x)(Fx \supset Gx)$, all ten tosses will be heads. Our new conjecture is that there is a physical asymmetry in the coin, an asymmetry that tends to yield more heads than tails. We have succeeded in confirming the general hypothesis (all ten tosses will be heads), but we have done so via an intermediate hypothesis involving *genuine laws* relating the physical make-up of the coin to the frequency of heads in a population of tosses.

It is by such devices as this that we create for ourselves, or some philosophers create for themselves, the *illusion* that (apart from supplementary *law-like* assumptions) general truths can be confirmed by their instances and therefore qualify, in this respect, as laws of nature. The illusion is fostered in the following way. It is assumed that confirmation is a matter of *raising the probability of a hypothesis.*[10] On this assumption any general statement of finite scope can be confirmed by examining its instances and finding them favorable. The hypothesis about the results of flipping a coin ten times can be confirmed by tossing nine straight heads, and this confirmation takes place without *any* assumptions about the coin's bias. Similarly, I confirm (to some degree) the hypothesis that all the people in the hotel ballroom are over thirty years old when I enter the ballroom with my wife and realize that *we* are both over thirty. In both cases I raise the probability that the hypothesis is true over what it was originally (before flipping the coin and before entering the ballroom). But this, of course, isn't confirmation. Confirmation is not simply raising the probability that a hypothesis is true, it is raising the probability that the unexamined cases resemble (in the relevant respect) the examined cases. It is *this* probability that must be raised if genuine confirmation is to occur (and if a confirmed hypothesis to be useful in *prediction*), and it is precisely this probability that is left unaffected by the instantial "evidence" in the above examples.

In order to meet this difficulty, and to cope with hypotheses that are *not* of limited scope,[11] the reductionist usually smuggles into his confirmatory proceedings the very idea he professes to do without: viz., a type of law that is not merely a universal truth. The general truth then gets confirmed but *only* through the mediation of these supplementary laws. These supplementary assumptions are usually introduced to *explain* the regularities manifested in the examined instances so as to provide a basis for projecting these regularities to the unexamined cases. The only way we can get a purchase on the unexamined cases is to introduce a hypothesis which, while *explaining* the data we already have, *implies* something about the data we do not have. To suppose that our coin is biased (first example) is to suppose something that contributes to the explanation of our extraordinary run of heads (nine straight) and simultaneously implies something about the (probable) outcome of the tenth toss. Similarly (second example) my wife and I may be attending a reunion of some kind, and I may suppose that the other people in the ballroom are old classmates. This hypothesis not only explains our presence, it implies that most, if not all, of the remaining people in the room are of comparable age (well over thirty). In both these cases the generalization can be confirmed, but only via the introduction of a law or circumstance (combined with a law or laws) that helps to explain the data already available.

One additional example should help to clarify these last remarks. In sampling from an urn with a population of colored marbles, I can confirm the hypothesis that all the marbles in the urn are red by extracting at random several dozen red marbles (and no marbles of any other color). This is a genuine example of confirmation, not because I have raised the probability of the hypothesis that all are red by reducing the number of ways it can be false (the same reduction would be achieved if you *showed* me 24 marbles from the urn, all of which were red), but because the hypothesis that all the marbles in the urn are red, together with the fact (law) that you cannot draw nonred marbles from an urn containing only red marbles, *explains* the result of my random sampling. Or, if this is too strong, the law that assures me that random sampling from an urn containing a substantial number of nonred marbles would reveal (in all likelihood) at least one nonred marble lends its support to my confirmation that the urn contains only (or mostly) red marbles. Without the assistance of such auxiliary laws a sample of 24 red marbles is powerless to confirm a hypothesis about the total population of marbles in the urn. To suppose otherwise is to suppose that the *same* degree of confirmation would be afforded the hypothesis if you, whatever your deceitful intentions, showed me a carefully selected set of 24 red marbles from the urn. This *also* raises the probability that they are all red, but the trouble is that it does not (due to your unknown motives and intentions) raise the probability that the unexamined marbles resemble the examined ones. And it does not raise this probability because we no longer have, as the best available explanation of the examined cases (all red), a hypothesis that implies that the remaining (or most of the remaining) marbles are also red. Your careful selection of 24 red marbles from an urn containing many different colored marbles is an equally good explanation of the data and it does not imply that the remainder are red. Hence, it is not just the fact that we have 24 red marbles in our sample class (24 positive instances and no negative instances) that confirms the general hypothesis that all the marbles in the urn are red. It is this data *together with a law* that confirms it, a law that (together with the hypothesis) explains the data in a way that the general hypothesis alone cannot do.

We have now reached a critical stage in our examination of the view that a properly qualified set of universal generalizations can serve as the fundamental laws of nature. For we have, in the past few paragraphs, introduced the notion of *explanation,* and it is this notion, perhaps more than any other, that has received the greatest attention from philosophers in their quest for the appropriate X in the formula: law = universal truth + X. R. B. Braithwaite's treatment ([3]) is typical. He begins by suggesting that it is merely deductive integration that transforms a universal truth into a law of nature. Laws are simply universally true statements of the form $(x)(Fx \supset Gx)$ that are derivable from certain higher level hypotheses. To say that $(x)(Fx \supset Gx)$ is a statement of law is to say, not only that it is true, but that it is *deducible from* a higher level hypothesis, H, in a well-established scientific system. The fact that it must be deducible from some higher level hypothesis, H, confers on the statement the opacity we are seeking to understand. For we may have a hypothesis from which we can derive $(x)(Fx \supset Gx)$ but from which we cannot derive $(x)(Kx \supset Gx)$ despite the coextensionality of *"F"* and *"K."* Braithwaite also argues that such a view gives a satisfactory account of the counterfactual force of laws.

The difficulty with this approach (a difficulty that Braithwaite recognizes) is that it only postpones the problem. Something is not a statement of law simply because it is true and deducible from some well-established higher level hypothesis. For every generalization implies another of smaller scope (e.g., $(x)(Fx \supset Gx)$ implies $(x)(Fx \,\&\, Hx \supset Gx)$), but this fact has not the slightest tendency to transform the latter generalization into a law. What is required is that the higher level hypothesis *itself* be law-like. You cannot give to others what you do not have yourself. But now, it seems, we are back where we started. It is at this point that Braithwaite begins talking about the higher level hypotheses having *explanatory force* with respect to the hypotheses subsumed under them. He is forced into this maneuver to account for the fact that these higher level hypotheses—not themselves law-like on his characterization (since not themselves derivable from still higher level hypotheses)—are capable of conferring law-likeness on their consequences. The higher level hypotheses are laws because they explain; the lower level hypotheses are laws because they are deducible from laws. This fancy twist smacks of circularity. Nevertheless, it represents a conversion to *explanation* (instead of *deducibility)* as the fundamental feature of laws, and Braithwaite concedes this: "A hypothesis to be regarded as a natural law must be a general proposition which can be thought to *explain* its instances" ([3], p. 303) and, a few lines later, "Generally speaking, however, a true scientific hypothesis will be regarded as a law of nature if it has an explanatory function with regard to lower-level hypotheses or its instances." Deducibility is set aside as an incidental (but, on a Hempelian model of explanation, an important) facet of the more ultimate idea of explanation.

There is an added attraction to this suggestion. As argued above, it is difficult to see how instantial evidence can serve to confirm a universal generalization of the form: $(x)(Fx \supset Gx)$. If the generalization has an infinite scope, the ratio "examined favorable cases/total number of cases" never increases. If the generalization has a finite scope, or we treat its probability as something other than the above ratio, we may succeed in raising its probability by finite samples, but it is never clear how we succeed in raising the probability that the unexamined cases resemble the examined cases without invoking laws as auxiliary assumptions. And this is the very notion we are trying to analyze. To this problem the notion of explanation seems to provide an elegant rescue. If laws are those universal generalizations that explain their instances, then following the lead of a number of current authors (notably Harman ([8], [9]); also see Brody ([4])), we may suppose that universal generalizations can be confirmed because confirmation is (roughly) the converse of explanation; *E* confirms *H* if *H* explains *E*. *Some* universal generalizations can be confirmed; they are those that explain their instances. Equating laws with universal generalizations having explanatory power therefore achieves a neat economy: we account for the confirmability of laws in terms of the explanatory power of those generalizations to which laws are reduced.

To say that a law is a universal truth having explanatory power is like saying that a chair is a breath of air used to seat people. You cannot make a silk purse out of a sow's ear, not even a very good sow's ear; and you cannot *make* a generalization, not even a purely universal generalization, explain its instances. The fact that *every F* is *G* fails to explain why *any F* is *G*, and it fails to explain it, not because its explanatory efforts are too feeble to have attracted our attention, but because the explanatory attempt is never even made. The fact that all men are mortal does not explain why you and I are mortal; it *says* (in the sense of *implies*) that we are mortal, but it does not even suggest *why* this might be so. The fact that all ten tosses will turn up heads is a fact that logically guarantees a head on the tenth toss, but it is not a fact that explains the outcome of this final toss. On one view of explanation, *nothing* explains it. Subsuming an instance under a universal generalization has exactly as much explanatory power as deriving Q from P & Q. None.

If universal truths of the form $(x)(Fx \supset Gx)$ could be *made* to explain their instances, we might succeed in making them into natural laws. But, as far as I can tell, no one has yet revealed the secret for endowing them with this remarkable power.

This has been a hasty and, in some respects, superficial review of the doctrine that laws are universal truths. Despite its brevity, I think we have touched upon the major difficulties with sustaining the equation: law = universal truth + X (for a variety of different values of "X"). The problems center on the following features of laws:

(a) A statement of law has its descriptive terms occurring in opaque positions.

(b) The existence of laws does not await our identification of them *as* laws. In this sense they are objective and independent of epistemic considerations.

(c) Laws can be confirmed by their instances and the confirmation of a law raises the probability that the unexamined instances will resemble (in the respect described by the law) the examined instances. In this respect they are useful tools for prediction.

(d) Laws are not merely summaries of their instances; typically, they figure in the explanation of the phenomena falling within their scope.

(e) Laws (in some sense) "support" counterfactuals; to know a law is to know what would happen if certain conditions were realized.

(f) Laws tell us what (in some sense) must happen, not merely what has and will happen (given certain initial conditions).

The conception of laws suggested earlier in this essay, the view that laws are expressed by singular statements of fact describing the relationships between properties and magnitudes, proposes to account for these features of laws in a single, unified way: (a)–(f) are all manifestations of what might be called "ontological ascent," the shift from talking about individual objects and events, or collections of them, to the quantities and qualities that these objects exemplify. Instead of talking about green and red things, we talk about the *colors* green and blue. Instead of talking about gases that have a volume, we talk about the volume (temperature, pressure, entropy) that gases have. Laws eschew reference to the things that have length, charge, capacity, internal energy, momentum, spin, and velocity in order to talk about these quantities themselves and to describe *their* relationship to each other.

We have already seen how this conception of laws explains the peculiar opacity of law-like statements. Once we understand that a law-like statement is not a statement about the extensions of its constituent terms, but about the intensions (= the quantities and qualities to which we may refer with the abstract singular form of these terms), then the opacity of laws to *extensional* substitution is natural and expected. Once a law is understood to have the form:

(6) F-ness $\rightarrow$ G-ness

the relation in question (the relation expressed by "$\rightarrow$") is seen to be an *extensional* relation between *properties* with the terms "F-ness" and "G-ness" occupying *transparent* positions in (6). Any term referring to the same quality or quantity as "F-ness" can be substituted for "F-ness" in (6) without affecting its truth or its law-likeness. Coextensive terms (terms referring to the same *quantities and qualities*) can be freely exchanged for "F-ness" and "G-ness" in (6) without jeopardizing its truth value. The tendency to treat laws as some kind of intensional relation between extensions, as something of the form $(x)(Fx\,\boxed{\text{n}}\!\rightarrow Gx)$ (where the connective is some kind of modal connective), is simply a mistaken rendition of the fact that laws are extensional relations between intensions.

Once we make the ontological ascent we can also understand the modal character of laws, the feature described in (e) and (f) above. Although true statements having the form of (6) are not themselves *necessary* truths, nor do they describe a modal relationship between the respective qualities, the contingent relationship between properties that is described imposes a modal quality on the particular events falling within its scope. This F *must* be G. Why? Because F-ness is linked to G-ness; the one property yields or generates the other in much the way a change in the thermal conductivity of a metal yields a change in its electrical conductivity. The pattern of inference is:

(I) F-ness $\rightarrow$ G-ness

This is F

This must be G.

This, I suggest, is a valid pattern of inference. It is quite unlike the fallacy committed in (II):

(II) $(x)(Fx \supset Gx)$.

This is F

This must be *G*.

The fallacy here consists in the absorption *into* the conclusion of a modality (entailment) that belongs to the relationship *between* the premises and the conclusion. There is no fallacy in (I), and this, I submit, is the source of the "physical" or "nomic" necessity generated by laws. It is this which explains the power of laws to tell us what *would* happen if we did such-and-such and what *could not* happen whatever we did.

I have no proof for the validity of (I). The best I can do is an analogy. Consider the complex set of legal relationships defining the authority, responsibilities, and powers of the three branches of government in the United States. The executive, the legislative, and the judicial branches of government have, according to these laws, different functions and powers. There is nothing *necessary* about the laws themselves; they could be changed. There is no law that prohibits scrapping all the present laws (including the constitution) and starting over again. Yet, given these laws, it follows that the president *must* consult Congress on certain matters, members of the Supreme Court *cannot* enact laws nor declare war, and members of Congress *must* periodically stand for election. The legal code lays down a set of relationships between the various *offices* of government, and this set of relationships (between the abstract offices) imposes legal constraints on the individuals who occupy these offices—constraints that we express with such modal terms as "cannot" and "must." There are certain things the individuals (and collections of individuals—e.g., the Senate) can and cannot do. *Their* activities are subjected to this modal qualification whereas the framework of laws from which this modality arises is itself modality-free. The president (e.g., Ford) *must* consult the Senate on matter *M*, but the relationship between the *office* of the president and that *legislative body* we call the Senate that makes Gerald Ford's action obligatory is not *itself* obligatory. There is no law that says that this relationship between the office of president and the upper house of Congress must (legally) endure forever and remain indisoluble.

In matters pertaining to the offices, branches, and agencies of government the "can" and "cannot" generated by laws are, of course, legal in character. Nevertheless, I think the analogy is revealing. Natural laws may be thought of as a set of relationships that exist between the various "offices" that objects sometimes occupy. Once an object occupies such an office, its activities are constrained by the set of relations connecting that office to other offices and agencies: it *must* do some things, and it *cannot* do other things. In both the legal and the natural context the modality at level n is generated by the set of relationships existing between the entities at level $n + 1$. Without this web of higher order relationships there is nothing to support the attribution of constraints to the entities at a lower level.

To think of statements of law as expressing relationships (such as class inclusion) between the extensions of their terms is like thinking of the legal code as a set of universal imperatives directed to a set of particular individuals. A law that tells us that the United States president must consult Congress on matters pertaining to *M* is not an imperative issued to Gerald Ford, Richard Nixon, Lyndon Johnson et al. The law tells us something about the duties and obligations attending the *presidency;* only indirectly does it tell us about the obligations of the presidents (Gerald Ford, Richard Nixon et al.). It tells us about their obligations in so far as they are occupants of this office. If a law was to be interpreted as of the form: "For all *x,* if *x* is (was or will be) president of the United States, then *x* must (legally) consult Congress on matter *M,*" it would be incomprehensible why Sally Bickle, were she to be president, would have to consult Congress on matter *M.* For since Sally Bickle never was and never will be president, the law, understood as an imperative applying to *actual* presidents (past, present, and future) does not apply to her. Even if there is a possible world in which she becomes president, this does not make her a member of that class of people to which the law applies; for the law, under this interpretation, is directed to that class of people who become president in *this* world, and Sally is not a member of this class. But we all know, of course, that the law does not apply to individuals, or sets of individuals, in this way; it concerns itself, in part, with the offices that people occupy and only indirectly with individuals insofar as they occupy these offices. And this is why, if Sally Bickle were to become president, if she occupied this office, she would have to consult Congress on matters pertaining to *M.*[12]

The last point is meant to illustrate the respect and manner in which natural laws "support" counterfactuals. Laws, being relationships between properties and magnitudes, *go beyond* the sets of things in *this* world that exemplify these properties and have these magnitudes. Laws tell us that quality *F* is linked to quality *G* in a certain way; hence, if object *O* (which has neither property) were to acquire property *F,* it would also acquire *G* in virtue of this connection between *F*-ness and *G*-ness. A statement of law asserts something that allows us to entertain the prospect of alterations in the extension of the predicate expressions contained in the statement. Since they make no reference to the extensions of their constituent terms (where the extensions are understood to be the things that are *F* and *G* in this world), we can hypothetically alter these extensions in the antecedent of our counterfactual ("if this were an *F*. . .") and use the connection asserted in the law to reach the consequent (". . . it would be *G*"). Statements of law, by talking about the relevant properties rather than the sets of things that have these properties, have a far wider scope than any true generalization about the actual world. Their scope extends to those possible worlds in which the extensions of our terms differ but the connections between properties remains invariant. This is a power that no universal generalization of the form $(x)(Fx \supset Gx)$ has; this statement says something about the actual *F*'s and *G*'s in *this* world. It says absolutely nothing about those possible worlds in which there are *additional F*'s or *different F*'s. For this reason it cannot imply a counterfactual. To do this we must ascend to a level of discourse in which what we talk about, and what we say about what we talk about, remains the *same* through alternations in extension. This can only be achieved through an ontological ascent of the type reflected in (6).

We come, finally, to the notion of explanation and confirmation. I shall have relatively little to say about these ideas, not because I think that the present conception of laws is particularly weak in this regard, but because its very real strengths have already been made evident. Laws figure in the explanation of their instances because they are not merely summaries of these instances. I can explain why this *F* is *G* by describing the relationship that exists between the properties in question. I can explain why the current increased upon an increase in the voltage by appealing to the relationship that exists between the flow of charge (current intensity) and the voltage (notice the definite articles). The period of a pendulum decreases when you shorten the length of the bob, not because all pendulums do that, but because the period and the length are related in the fashion $T = 2\pi\sqrt{L/g}$. The principles of thermodynamics tell us about the relationships that exist between such quantities as energy, entropy, temperature, and pressure, and it is for this reason that we can use these principles to explain the increase in temperature of a rapidly compressed gas, explain why perpetual motion machines cannot be built, and why balloons do not spontaneously collapse without a puncture.

Furthermore, if we take seriously the connection between explanation and confirmation, take seriously the idea that to confirm a hypothesis is to bring forward data for which the hypothesis is the best (or one of the better) competing explanations, then we arrive at the mildly paradoxical result that laws can be confirmed *because* they are more than generalizations of that data. Recall, we began this essay by saying that if a statement of law asserted anything more than is asserted by a universally true statement of the form $(x)(Fx \supset Gx)$, then it asserted something that was beyond our epistemological grasp. The conclusion we have reached is that *unless* a statement of law goes beyond what is asserted by such universal truths, unless it asserts something that cannot be completely verified (even with a complete enumeration of its instances), it cannot be confirmed and used for predictive purposes. It cannot be confirmed because it cannot explain; and its inability to explain is a symptom of the fact that there is not enough "distance" between it and the facts it is called upon to explain. To get this distance we require an ontological ascent.

I expect to hear charges of Platonism. They would be premature. I have not argued that there are universal properties. I have been concerned to establish something weaker, something conditional in nature: viz., universal properties exist, and there exists a definite relationship between these universal properties, if there are any laws of nature. If one prefers desert landscapes, prefers to keep one's ontology respectably nominalistic, I can and do sympathize. I would merely point out that in such barren terrain there are no laws, nor is there anything that can be dressed up to look like a law. These are inflationary times, and the cost of nominalism has just gone up.

Notes

For their helpful comments my thanks to colleagues at Wisconsin and a number of other universities where I read earlier versions of this paper. I wish, especially, to thank Zane Parks, Robert Causey, Martin Perlmutter, Norman Gillespie, and Richard Aquilla for their critical suggestions, but they should not be blamed for the way I garbled them.

1. This is the position taken by Hempel and Oppenheim ([10]).

2. When the statement is of nonlimited scope it is already beyond our epistemological grasp in the sense that we cannot *conclusively* verify it with the (necessarily) finite set of

observations to which traditional theories of confirmation restrict themselves. When I say (in the text) that the statement is "beyond our epistemological grasp" I have something more serious in mind than this rather trivial limitation.

3. Most prominently, William Kneale in [12] and [13].

4. I eliminate quotes when their absence will cause no confusion. I will also, sometimes, speak of laws and statements of law indifferently. I think, however, that it is a serious mistake to conflate these two notions. Laws are what is expressed by true law-like statements (see [1], p. 2, for a discussion of the possible senses of "law" in this regard). I will return to this point later.

5. Popper ([17]) vaguely perceives, but fails to appreciate the significance of, the same (or a similar) point. He distinguishes between the structure of terms in laws and universal generalizations, referring to their occurrence in laws as "intensional" and their occurrence in universal generalizations as "extensional." Popper fails to develop this insight, however, and continues to equate laws with a certain class of universal truths.

6. Nelson Goodman gives a succinct statement of the functionalist position: "As a first approximation then, we might say that a law is a true sentence used for making predictions. That laws are used predictively is of course a simple truism, and I am not proposing it as a novelty. I want only to emphasize the Humean idea that rather than a sentence being used for prediction because it is a law, it is called a law because it is used for prediction, and that rather than the law being used for prediction because it describes a causal connection, the meaning of the causal connection is to be interpreted in terms of predictively used laws" ([7], p. 26). Among functionalists of this sort I would include Ayer ([2]), Nagel ([16]), Popper ([17]), Mackie ([14]), Bromberger ([6]), Braithwaite ([3]), Hempel ([10], [11]), and many others. Achinstein is harder to classify. He says that laws express regularities that can be cited in providing analyses and explanations ([1], p. 9), but he has a rather broad idea of regularities: "regularities might also be attributed to properties" ([1]), pp. 19, 22).

7. I attach no special significance to the connective "$\rightarrow$." I use it here merely as a dummy connective or relation. The kind of connection asserted to exist between the universals in question will depend on the particular law in question, and it will vary depending on whether the law involves quantitative or merely qualitative expressions. For example, Ohm's Law asserts for a certain class of situations a constant ratio (R) between the magnitudes E (potential difference) and I (current intensity), a fact that we use the "=" sign to represent: $E/I = R$. In the case of simple qualitative laws (though I doubt whether there are many genuine laws of this sort) the connective "$\rightarrow$" merely expresses a link or connection between the respective qualities and may be read as "yields." If it is a law that all men are mortal, then humanity yields mortality (humanity $\rightarrow$ mortality). Incidentally, I am not denying that we can, and do, express laws as simply "All *F*'s are *G*" (sometimes this is the only convenient way to express them). All I am suggesting is that when law-like statements are presented in this form it may not be clear what is being asserted: a law or a universal generalization. When the context makes it clear that a relation of law is being described, we can (without ambiguity) express it as "All *F*'s are *G*" for it is then understood in the manner of (6).

8. On the basis of an argument concerned with the restrictions on predicate expres-

sions that may appear in laws, Hempel reaches a similar conclusion but he interprets it differently. "Epitomizing these observations we might say that a law-like sentence of universal nonprobabilistic character is not about classes or about the extensions of the predicate expressions it contains, but about these classes or extensions *under certain* descriptions" ([11], p. 128). I guess I do not know what being *about* something *under a description* means unless it amounts to being about the property or feature expressed by that description. I return to this point later.

9. Molnar ([15]) has an excellent brief critique of attempts to analyze a law by using epistemic conditions of the kind being discussed.

10. Brody argues that a qualitative confirmation function need not require that any E that raises the degree of confirmation of H thereby (qualitatively) confirms H. We need only require (perhaps this is also too much) that if E does qualitatively confirm H, then E raises the degree of confirmation of H. His arguments take their point of departure from Carnap's examples against the special consequence and converse consequence condition ([4], pp. 414–18). However this may be, I think it fair to say that most writers on confirmation theory take a *confirmatory* piece of evidence to be a piece of evidence that *raises* the probability of the hypothesis for which it is confirmatory. How well it must be confirmed to be acceptable is another matter of course.

11. If the hypothesis is of nonlimited scope, then its scope is not known to be finite. Hence, we cannot know whether we are getting a numerical increase in the ratio: examined favorable cases/total number of cases. If an increase in the probability of a hypothesis is equated with a (known) increase in this ratio, then we cannot raise the probability of a hypothesis of nonlimited scope in the simple-minded way described for hypotheses of (known) finite scope.

12. If the law was interpreted as a universal imperative of the form described, the most that it would permit us to infer about Sally would be a counteridentical: If Sally were one of the presidents (i.e., identical with either Ford, Nixon, Johnson, . . .), then she would (at the appropriate time) have to consult Congress on matters pertaining to M.

References

[1] Achinstein, P. *Law and Explanation.* Oxford: Clarendon Press, 1971.

[2] Ayer, A. J. "What Is a Law of Nature." In [5], pp. 39–54.

[3] Braithwaite, R. B. *Scientific Explanation.* Cambridge, England: Cambridge University Press, 1957.

[4] Brody, B. A. "Confirmation and Explanation." *Journal of Philosophy* 65 (1968): 282–99. Reprinted in [5], pp. 410–26.

[5] Brody, B. A. *Readings in the Philosophy of Science.* Englewood Cliffs, N.J.: Prentice Hall, 1970.

[6] Bromberger, S. "Why-Questions." In [5], pp. 66–87.

[7] Goodman. N. *Fact, Fiction and Forecast.* London: The Athlone Press, 1954.

[8] Harman, G. "The Inference to the Best Explanation." *Philosophical Review* 74 (1965): 88–95.

[9] Harman, G. "Knowledge, Inference and Explanation." *Philosophical Quarterly* 18 (1968): 164–73.

[10] Hempel, C. G., and Oppenheim, P. "Studies in the Logic of Explanation." In [5], pp. 8–27.

[11] Hempel, C. G. "Maximal Specificity and Lawlikeness in Probabilistic Explanations." *Philosophy of Science* 35 (1968): 116–33.

[12] Kneale, W. "Natural Laws and Contrary-to-Fact Conditionals." *Analysis* 10 (1950): 121–25.

[13] Kneale, W. *Probability and Induction.* Oxford: Oxford University Press, 1949.

[14] Mackie, J. L. "Counterfactuals and Causal Laws." In *Analytical Philosophy.* (First Series). Edited by R. J. Butler. Oxford: Basil Blackwell, 1966.

[15] Molnar. G. "Kneale's Argument Revisited." *Philosophical Review* 78 (1969): 79–89.

[16] Nagel, E. *The Structure of Science.* New York: Harcourt Brace, 1961.

[17] Popper, K. "A Note on Natural Laws and So-Called 'Contrary-to-Fact Conditionals.'" *Mind* 58 (1949): 62–66.

2

The Nature of Laws

MICHAEL TOOLEY

This paper is concerned with the question of the truth conditions of nomological statements. My fundamental thesis is that it is possible to set out an acceptable, noncircular account of the truth conditions of laws and nomological statements if and only if relations among universals—that is, among properties and relations, construed realistically—are taken as the truth-makers for such statements.

My discussion will be restricted to strictly universal, nonstatistical laws. The reason for this limitation is not that I feel there is anything dubious about the concept of a statistical law, nor that I feel that basic laws cannot be statistical. The reason is methodological. The case of strictly universal, nonstatistical laws would seem to be the simplest case. If the problem of the truth conditions of laws can be solved for this simple subcase, one can then investigate whether the solution can be extended to the more complex cases. I believe that the solution I propose here does have that property, though I shall not pursue that question here.

1. Some Unsatisfactory Accounts

The thesis that relations among universals are the truth-makers for laws may strike some philosophers as rather unappealing, for a variety of reasons. Perhaps the two most important are these. First, it entails a strong version of realism with regard to universals. Secondly, traditional semantical accounts of the concept of truth have generally been nominalistic in flavor. Not in the sense that acceptance of them involves commitment to nominalism, but in the sense that they involve no reference to universals. This seems, in part, an historical accident. Semantical accounts of truth in

This essay is reprinted with permission of the University of Calgary Press from *Canadian Journal of Philosophy* 7 (1977): 667–98.

which a concept such as an object's exemplifying a property plays a central role can certainly be set out. For most types of sentences, accounts involving such explicit reference to universals may well introduce additional conceptual apparatus without any gain in philosophical illumination. However I will attempt to show that there is at least one class of statements for which this is not the case, namely, nomological statements.

I shall begin by considering some important alternative accounts of the nature of laws. I think that getting clear about how these accounts are defective will both point to certain conditions that any adequate account must satisfy, and provide strong support for the thesis that the truth-makers for laws must be relations among universals.

Perhaps the most popular account of the nature of laws is that a generalization expresses a law if and only if it is both lawlike and true, where lawlikeness is a property that a statement has, or lacks, simply in virtue of its meaning. Different accounts of lawlikeness have been advanced, but one requirement is invariably taken to be essential: a lawlike statement cannot contain any essential reference to specific individuals. Consider, for example, the generalization: "All the fruit in Smith's garden are apples." Since this statement entails the existence of a particular object—Smith's garden—it lacks the property of lawlikeness. So unless it is entailed by other true statements which are lawlike, it will be at best an accidentally true generalization.

There are at least three serious objections to this approach. First, consider the statement that all the fruit in any garden with property *P* are apples. This generalization is free of all essential reference to specific individuals. Thus, unless it is unsatisfactory in some other way, it is lawlike. But *P* may be quite a complex property, so chosen that there is, as a matter of fact, only one garden possessing that property, namely Smith's. If that were so, one might well want to question whether the generalization that all fruit in any garden with property *P* are apples was a law. It would seem that statements can be both lawlike and true, yet fail to be laws.

A second objection to this approach is that it cannot deal in a satisfactory manner with generalizations that are vacuously true; that is, which lack "positive" confirming instances.[1] Consider the statement: "Whenever two spheres of gold more than eight miles in diameter come into contact, they turn red." The statement is presumably lawlike, and is true under the standard interpretation. Is it then a law? The usual response is that a vacuously true generalization is a law only if it is derivable from generalizations that are not vacuously true. But this seems wrong. Imagine a world containing ten different types of fundamental particles. Suppose further that the behavior of particles in interactions depends upon the types of the interacting particles. Considering only interactions involving two particles, there are 55 possibilities with respect to the types of the two particles. Suppose that 54 of these possible interactions have been carefully studied, with the result that 54 laws have been discovered, one for each case, which are not interrelated in any way. Suppose finally that the world is sufficiently deterministic that, given the way particles of types *X* and *Y* are currently distributed, it is impossible for them ever to interact at any time, past, present, or future. In such a situation it would seem very reasonable to believe that there is some underived law dealing with the interaction of particles of types *X* and *Y*. Yet precisely this view would have to be rejected if one were to accept the claim that a vacuously true generalization can be a law only if derivable from laws that are not vacuously true.

A third objection is this. Assuming that there can be statistical laws, let us suppose that it is a law that the probability that something with property *P* has property *Q* is 0.999999999. Suppose further that there are, as a matter of fact, very few things in the world with property *P* and, as would then be expected, it happens that all of these things have property *Q*. Then the statement that everything with property *P* has property *Q* would be both lawlike and true, yet it would not be a law.

One might even have excellent grounds for holding that it was not a law. There might be some powerful and very well-established theory which entailed that the probability that something with property *P* would have property *Q* was not 1.0, but 0.999999999, thus implying that it was not a law that everything with property *P* would have property *Q*.

If this argument is correct, it shows something quite important. Namely, that there are statements that would be laws in some worlds, but only accidentally true generalizations in others. So there cannot be any property of lawlikeness which a statement has simply in virtue of its meaning, and which together with truth is sufficient to make a statement a law.

A second attempt to explain what it is for a statement to express a law appeals to the fact that laws entail some counterfactuals, and support others, while accidentally true generalizations do neither. If this approach is to provide a noncircular analysis, it must be possible to give a satisfactory account of the truth conditions of subjunctive conditional statements which does not involve the concept of a law. This does not seem possible. The traditional, consequence analysis of subjunctive conditionals explicitly employs the concept of a law. And the principal alternative, according to which truth conditions for subjunctive conditionals are formulated in terms of comparative similarity relations among possible worlds, involves implicit reference to laws, since possession of the same laws is one of the factors that weighs most heavily in judgments concerning the similarity of different possible worlds. The latter theory is also exposed to very serious objections.[2] As a result, it appears unlikely that any noncircular analysis of the concept of a law in terms of subjunctive conditional statements is possible.

A third approach to the problem of analyzing the concept of a law is the view, advanced by Ramsey, that laws are "consequences of those propositions which we should take as axioms if we knew everything and organized it as simply as possible in a deductive system."[3] My earlier example of the universe in which there are ten different types of fundamental particles, two of which never interact, shows that this account does not provide an adequate description of the truth conditions of laws. In the world where particles of types *X* and *Y* never meet, there will be many true generalizations about their behavior when they interact. Unfortunately, none of these generalizations will have any positive instances; they will all be only vacuously true. So knowledge of everything that happens in such a universe will not enable one to formulate a *unique* axiomatic system containing theorems about the manner of interaction of particles of types *X* and *Y*. Adopting Ramsey's approach would force one to say that in such a universe there could not be any law describing how particles of types *X* and *Y* would behave if they were to interact. I have argued that this is unacceptable.

2. Universals as the Truth-Makers for Nomological Statements

What, then, is it that makes a generalization a law? I want to suggest that a fruitful place to begin is with the possibility of underived law having no positive instances. This possibility brings the question of what makes a generalization a law into very sharp focus, and it shows that an answer that might initially seem somewhat metaphysical is not only plausible, but unavoidable.

Consider, then, the universe containing two types of particles that never meet. What in that world could possibly make true some specific law concerning the interaction of particles of types *X* and *Y*? All the events that constitute the universe throughout all time are perfectly compatible with different, and conflicting, laws concerning the interaction of these two types of particles. At this point one may begin to feel the pull of the view that laws are not statements, but inference tickets. For in the universe envisaged, there is nothing informative that one would be justified in inferring from the supposition that an *X* type particle has interacted with a *Y* type particle. So if laws are inference tickets, there are, in our imaginary universe, no laws dealing with the interaction of particles of types *X* and *Y*.

But what if, in the universe envisaged, there could be underived laws dealing with the interaction of particles of types *X* and *Y?* Can one draw any conclusions from the assumption that such basic laws without positive instances are possible—specifically, conclusions about the truth-makers for laws? I would suggest that there are two very plausible conclusions. First, *nonnomological* facts about particulars cannot serve as the truth-makers for *all* laws. In the universe in which particles of types *X* and *Y* never interact, it might be a law that when they do, an event of type *P* occurs. But equally, it might be a law that an event of type *Q* occurs. These two generalizations will not be without instances, but none of them will be of the positive variety. And in the absence of positive instances, there is no basis for holding that one generalization is a law, and the other not. So at least in the case of underived laws without positive instances, nonnomological facts about particulars cannot serve as the truth-makers.

What, then, are the facts about the world that make such statements laws? A possible answer is that the truth-makers are facts about particulars that can only be expressed in *nomological* language. Thus, in the case we are considering, one might try saying that what makes it a law that particles of types *X* and *Y* interact in a specific way are the nomological facts that particles of types *X* and *Y* have certain dispositional properties. But this is not to make any progress. The question of the truth-makers of underived laws without positive instances has merely been replaced by that of the truth-makers of statements attributing unactualized dispositional properties to objects, and if one is willing in the latter case to say that such statements are semantically basic, and that no further account can be given of the fact that an object has a dispositional property, one might equally well say the same thing of laws, that is, that there just are basic facts to the effect that there are specific laws applying to certain types of objects, and no further account of this can be given. In either case one is abandoning the project of providing an account of the truth conditions of nomological statements in nonnomological terms, and thus also the more general program of providing truth conditions for intensional statements in purely extensional terms.

The upshot, then, is that an account of the truth conditions of underived laws without positive instances in terms of nomological facts about particulars is unilluminating, while an account in terms of nonnomological facts about particulars seems impossible. This, then, is the second conclusion: no facts about particulars can provide a satisfactory account of the truth conditions of such laws.

But how then can there be such laws? The only possible answer would seem to be that it must be facts about *universals* that serve as the truth-makers for basic laws without positive instances. But if facts about universals constitute the truth-makers for some laws, why shouldn't they constitute the truth-makers for all laws? This would provide a uniform account of the truth conditions of laws, and one, moreover, that explains in a straightforward fashion the difference between laws and accidentally true generalizations.

Let us now consider how this idea that facts about universals can be the truth-makers for laws is to be developed. Facts about universals will consist of universals' having properties and standing in relations to other universals. How can such facts serve as truth-makers for laws? My basic suggestion here is that the fact that universals stand in certain relationships may *logically necessitate* some corresponding generalization about particulars, and that when this is the case, the generalization in question expresses a law.

This idea of a statement about particulars being entailed by a statement about a relation among universals is familiar enough in another context, since some philosophers have maintained that analytical statements are true in virtue of relations among universals. In this latter case, the relations must be necessary ones in order for the statement about particulars which is entailed to be itself logically necessary. Nomological statements, on the other hand, are not logically necessary, and because of this the relations among universals involved here must be *contingent* ones.

The idea of contingent relations among universals logically necessitating corresponding statements about particulars is admittedly less familiar. But why should it be more problematic? Given the relationship that exists between universals and particulars exemplifying them, any property of a universal, or relation among universals, regardless of whether it is necessary or merely contingent, will be reflected in corresponding facts involving the particulars exemplifying the universal.

It might be suggested, though, that what is problematic is rather the idea of a contingent relation among universals. Perhaps this idea is, like the notion of a necessarily existent being, ultimately incoherent? This possibility certainly deserves to be examined. Ideally, one would like to be able to prove that the concept of contingent relations among universals is coherent. Nevertheless, one generally assumes that a concept is coherent unless there are definite grounds for thinking otherwise. So unless some reason can be offered for supposing that the concept of contingent relations among universals is incoherent, one would seem to be justified in assuming that this is not the case.

Let us refer to properties of universals, and relations among universals, as *nomological* if they are contingent properties or relations which logically necessitate corresponding facts about particulars. How can one *specify* such nomological properties and relations? If the properties or relations were observable ones, there would be no problem. But in our world, at least, the facts about universals which are the truth-makers for laws appear to be unobservable. One is dealing, then, with theoretical relations among universals, and the problem of specifying nomological relations and properties is just a special case of the problem of specifying the meaning of statements involving theoretical terms.

Theoretical statements cannot be analyzed in purely observational terms. From this, many have concluded that theoretical statements cannot, in the strict sense, be analyzed at all in terms of statements free of theoretical vocabulary. But it is clear that this does not follow, since the class of statements that are free of theoretical vocabulary does not coincide with the class of observation statements. Thus the statement, "This table has parts too small to be observed," although it contains no theoretical vocabulary, is not a pure observation statement since it refers to something beyond what is observable. This situation can arise because, in addition to observational vocabulary and theoretical vocabulary, one also has logical and quasi-logical vocabulary—including expressions such as "part," "property," "event," "state," "particular," and so on—and statements containing such vocabulary together with observational vocabulary can refer to unobservable states of affairs.

This suggests the possibility, which I believe to be correct, that theoretical statements, though not analyzable in terms of observation statements, are analyzable in terms of statements that contain nothing beyond observational, logical, and quasi-logical vocabulary. The natural and straightforward way of doing this is suggested by the method of Ramsey sentences, and has been carefully worked out and defended by David Lewis in his article, "How to Define Theoretical Terms."[4] Essentially, the idea is this. Let T be any theory. If T contains any singular theoretical terms, eliminate them by paraphrase. Then replace all theoretical predicates and functors by names of corresponding entities, so that, for example, an expression such as ". . . is a neutrino" is replaced by an expression such as ". . . has the property of neutrino-hood." The result can be represented by $T(P_1, P_2, \ldots, P_n)$, where each P_i is the theoretical name of some property or relation. All such theoretical names are then replaced by distinct variables, and the corresponding existential quantifiers prefixed to the formula. The resulting sentence—$\exists x_1 \exists x_2 \ldots \exists x_n\ T(x_1, x_2, \ldots, x_n)$—is a Ramsey sentence for the theory T. Suppose now that there is only one ordered n-tuple that satisfies $T(x_1, x_2, \ldots, x_n)$. It will then be possible to define the theoretical name P_i by identifying the property or relation in question with the i_{th} member of the unique n-tuple which satisfies $T(x_1, x_2, \ldots, x_n)$.

Expressed in intuitive terms, the underlying idea is this. The meaning of theoretical terms is to be analyzed by viewing them as referring to properties (or relations) by characterizing them as properties (or relations) that stand in certain logical or quasi-logical relations to other properties and relations, both theoretical and observable, the logical and quasi-logical relations being specified by the relevant theory. One might compare here the way in which the mind is characterized in central state materialism: the mind is that entity, or collection of states and processes, that stands in certain specified relations to behavior. The above approach to the meaning of theoretical statements involves, in effect, a similar relational analysis of theoretical terms.

Some possible objections to this approach deserve to be at least briefly noted. One is that the procedure presupposes that the theoretical names will not name anything unless there is a unique n-tuple that satisfies the appropriate formula. I think that Lewis makes out a plausible case for this view. However the requirement can be weakened slightly. One might, for example, take the view that P_i is definable even where there is more than one n-tuple that satisfies $T(x_1, x_2, \ldots, x_n)$, provided that every n-tuple has the same i_{th} element.

A second objection is that replacing predicates and functors by names is not automatic, as Lewis supposes. For unless there are disjunctive properties, negative properties, etc., there is no reason for thinking that there will be a one-to-one correspondence between predicates on the one hand, and properties and relations on the other. This point is surely correct. However it shows only that the initial paraphrase has to be carried out in a metaphysically more sophisticated way.

A slightly more serious difficulty becomes apparent if one considers some very attenuated theories. Suppose, for example, that *T* consists of a single statement $(x)(Mx \supset Px)$, where *M* is theoretical and *P* is observational. This theory will have the peculiarity that the corresponding Ramsey sentence is logically true,[5] and given Lewis's approach, this means that the property *M*-hood exists only if it is identical with *P*-hood.

One response is to refuse to count a set of sentences as a theory if, like *T* here, it has no observational consequences. However this requirement seems overly stringent. Even if a theory does not entail any observational statements, it may have probabilistic implications: the likelihood of *R* given *Q* together with *T* may differ from the likelihood of *R* given *Q* alone, where *Q* and *R* are observational statements, even though there are no observational statements entailed by *T*.

An alternative response is to adopt the view that the Ramsey sentence for a theory should be replaced by a slightly different sentence which asserts not merely that there is some ordered *n*-tuple $(x_1, x_2, \ldots, x_n)$ that satisfies the formula $T(x_1, x_2, \ldots, x_n)$, but that there is some *n*-tuple that satisfies the formula *and* whose existence is not entailed simply by the existence of the observable properties involved in the theory. Then, if there is a unique ordered *n*-tuple with those two properties, one can define the theoretical terms $P_1, P_2, \ldots, P_n$.

In any case, I believe that one is justified in thinking that difficulties such as the above can be dealt with, and I shall, in my attempt to state truth conditions for laws, assume that theoretical statements can be adequately analyzed along the Lewis-type lines outlined above.

There are three further ideas that are needed for my account of the truth conditions of nomological statements. But the account will be more perspicuous if the motivation underlying the introduction of these ideas is first made clear. This can perhaps best be done by outlining a simpler version of the basic account, and then considering some possible problems that it encounters.[6]

The simpler version can be seen as attempting to specify explicitly a small number of relations among universals that will serve as the truth-makers for all possible laws. One relation which seems clearly essential is that of *nomic necessitation*. This relation can be characterized—though *not defined*—as that relation which holds between two properties *P* and *Q* if and only if it is a law that for all *x*, if *x* has property *P* then *x* has property *Q*. Is this one relation of nomic necessitation sufficient to handle all possible laws? The answer to this depends in part upon certain metaphysical matters. Consider a law expressed by a statement of the form $(x)(Px \supset -Qx)$. If this type of law is to be handled via the relation of nomic necessitation, one has to say that the property *P* stands in the relation of nomic necessitation to the property of not having property *Q*, and this commits one to the existence of negative properties. Since negative properties are widely thought suspect, and with good reason, another relation has to be introduced to handle laws of this form: the relation of *nomic exclusion*.

Are these two relations jointly sufficient? Consider some problematic cases. First, laws of the form $(x)Mx$. Is it possible to state truth conditions for such laws in terms of the relations of nomic necessitation and nomic exclusion? Perhaps. A first try would be to treat its being a law that everything has property *M* as equivalent to its being true, of every property *P*, that it is a law that anything that has property *P* also has property *M*. But whether this will do depends upon certain issues about the existence of properties. If different properties would have existed if the world of particulars had been different in certain ways, the suggested analysis will not be adequate. One will have to say instead that its being a law that everything has property *M* is equivalent to its being a law that, for every property *P*, anything with property *P* has property *M*—in order to exclude the possibility of there being some property *Q*, not possessed by any object in the world as it actually is, but which is such that if an object had property *Q* it would lack property *M*. This revision, since it involves the occurrence of a universal quantifier ranging over properties within the scope of a nomological operator, means that laws apparently about particulars are being analyzed in terms of laws about universals. This, however, would not seem to be a decisive objection to this way of viewing laws expressed by statements of the form $(x)Mx$.

A much more serious objection concerns laws expressed by statements of the form $(x)[Px \supset (Qx \vee Rx)]$. If the world were partially indeterministic, there might well be laws that stated, for example, that if an object has property *P*, then it has either property *Q* or property *R*, and yet no laws that specified *which* of those properties an object will have on any given occasion. Can the truth conditions for laws of this form be expressed in terms of the relations of nomic necessitation and nomic exclusion? The answer depends on whether there are disjunctive properties; that is, on whether, if Q and *R* are properties, there is a property, *Q or R*, which is possessed by objects that have property Q and by objects that have property *R*. If, as many philosophers have maintained, there are no disjunctive properties then the relations of nomic necessitation and nomic exclusion will not suffice to provide truth conditions for laws of the form $(x)[Px \supset (Qx \vee Rx)]$.

A third case that poses difficulties concerns laws expressed by statements of the form $(x)(-Px \supset Qx)$. If negative properties are rejected, such laws cannot be handled in any immediate fashion by the relations of nomic necessitation and nomic exclusion. Nevertheless, this third case does not appear to raise any new issues. For if one can handle laws expressed by sentences of the form $(x)Mx$ in the way suggested above, one can rewrite laws of the form $(x)(-Px \supset Qx)$ in the form $(x)(Px \vee Qx)$, and then apply the method of analysis suggested for laws of the form $(x)Mx$. The result will be a law that is conditional in form, with a positive antecedent and a disjunctive consequent, which is the case just considered.

The conclusion seems to be this. The relation of nomic necessitation by itself does not provide a satisfactory account unless there are both negative and disjunctive properties. Supplementing it with the relation of nomic exclusion may allow one to dispense with negative properties, but not with disjunctive ones. It would seem best to try to set out a more general account that will allow one to avoid all dubious metaphysical assumptions. Let us now turn to such an account.

The first concept required is that of the *universals involved in a proposition*. This notion is a very intuitive one, though how best to explicate it is far from clear. One approach would be to attempt to show that propositions can be identified with set theoretical constructs out of universals. This treatment of propositions is not without its difficulties, but it is a reasonably natural one if propositions are viewed as nonlinguistic entities.

The second, and related, idea is that of the logical form or structure of a proposition. One can view this form as specified by a *construction function* which maps ordered *n*-tuples of universals into propositions. Thus one could have, for example, a construction function *K* such that *K*(redness, roundness) = the proposition that all red things are round. Conceived in the most general way, some construction functions will map ordered n-tuples of universals into propositions that involve as constituents universals not contained in the original n-tuple. Thus G could be a function so defined that G(property P) = the proposition that everything with property P is green. Other construction functions will map ordered n-tuples of universals into propositions which do not contain, as constituents, all the universals belonging to the n-tuple. H could be a function so defined that H(property P, property Q) = the proposition that everything has property Q. In order to capture the notion of logical form, one needs a narrower notion of construction function, namely, one in which something is a construction function if and only if it is a mapping from ordered n-tuples of universals into propositions that contain, as constituents, all and only those universals belonging to the ordered n-tuple. In this narrower sense, K is a construction function, but G and H are not.

The final idea required is that of a *universal being irreducibly of order k.* Properties of, and relations among, particulars are universals of order one. If nominalism is false, they are irreducibly so. A universal is of order two if it is a property of universals of order one, or a relation among things, some of which are universals of order one, and all of which are either universals of order one or particulars. It is irreducibly so if it cannot be analyzed in terms of universals of order one. And in general, a universal is of order $(k + 1)$ if it is a property of universals of order k, or a relation among things, some of which are universals of order k, and all of which are either particulars or universals of order k or less. It is irreducibly of order $(k + 1)$ if it cannot be analyzed in terms of particulars and universals of order k or less.

Given these notions, it is possible to explain the general concept of a nomological relation—which will include, but not be restricted to, the relations of nomic necessitation and nomic exclusion. Thus, as a first approximation:

> R is a *nomological relation* if and only if
>
> (1) R is an n-ary relation among universals;
>
> (2) R is irreducibly of order $(k + 1)$, where k is the order of the highest order type of element that can enter into relation R;
>
> 3) R is a contingent relation among universals, in the sense that there are universals $U_1, U_2, \ldots, U_n$ such that it is neither necessary that $R(U_1, U_2, \ldots, U_n)$ nor necessary that not $R(U_1, U_2, \ldots, U_n)$;

(4) there is a construction function K such that

(i) if $P_1, P_2, \ldots, P_n$ are either properties or relations, and of the appropriate types, then $K(P_1, P_2, \ldots, P_n)$ is a proposition about particulars, and

(ii) the proposition that $R(P_1, P_2, \ldots, P_n)$ logically entails the proposition which is the value of $K(P_1, P_2, \ldots, P_n)$.

This characterization of the theoretical concept of a nomological relation in logical and quasi-logical terms can in turn be used to state truth conditions for nomological statements:

S is a true nomological statement if and only if there exists a proposition p which is expressed by S, and there exists a nomological relation R and an associated construction function K, and universals $P_1, P_2, \ldots, P_n$ such that

(1) it is not logically necessary that p;
(2) the proposition that p is identical with the value of $K(P_1, P_2, \ldots, P_n)$;
(3) it is true that $R(P_1, P_2, \ldots, P_n)$;
(4) it is not logically necessary that $R(P_1, P_2, \ldots, P_n)$;
(5) the proposition that $R(P_1, P_2, \ldots, P_n)$ logically entails the proposition that p.

The basic idea, then, is that a statement expresses a nomological state of affairs if it is true in virtue of a contingent, nomological relation holding among universals. Different types of nomological relations are specified by different construction functions. A relation is a relation of nomic necessitation if it is of the type specified by the construction function which maps ordered couples (P, Q) of universals into propositions of the form $(x)(Px \supset Qx)$. It is a relation of nomic exclusion if it is of the type determined by the construction function mapping ordered couples (P, Q) of universals into propositions of the form $(x)(Px \supset \text{-}Qx)$.

It is critical to this account that a nomological relation be genuinely a relation among universals and nothing else, as contrasted, for example, with a relation that is apparently among universals, but which can be analyzed in terms of properties of, and relations among, particulars. Hence the requirement that a relation, to be nomological, always be *irreducibly* of an order greater than the order of the universals that enter into it. If this requirement were not imposed, every true generalization would get classified as nomological. For suppose that everything with property P just happens to have property Q, and consider the relation R which holds between properties A and B if and only if everything with property A has property B. Properties P and Q stand in relation R. The relation is a contingent one. Its holding between properties P and Q entails the proposition that everything with property P has property Q. So if condition (2) were dropped from the definition of the concept of a nomological relation, R would qualify as a nomological relation, and it would be a nomological truth that everything with property P has property Q.

But while condition (2) is essential, it is not quite adequate. For suppose that it is a law that everything with property S has property T, and that the truth-maker for this law is the fact that S and T stand in a certain relation W, where W is irreducibly of order two. Then one can define a relation R as follows: Properties P and Q stand in relation R if and only if everything with property P has property Q, and properties S and T stand in relation W. So defined, relation R will not be analyzable in terms of universals of order one, so condition (2) will not be violated. But if relations such as R were admitted as nomological, then, provided that there was at least one true nomological statement, all generalizations about particulars would get classified as nomological statements.

There are alternative ways of coping with this difficulty. One is to replace condition (2) by:

> (2*) If $R(U_1, U_2, \ldots, U_n)$ is analytically equivalent to $C_1 \wedge C_2 \wedge \ldots \wedge C_m$, then every nonredundant C_i—that is, every C_i not entailed by the remainder of the conjunctive formula—is irreducibly of order $(k + 1)$.

This condition blocks the above counterexample. But given the somewhat ad hoc way in which it does so, one might wonder whether there may not be related counterexamples which it fails to exclude. What is one to say, for example, about a relation R defined as follows: Properties P and Q stand in relation R if and only if either everything with property P has property Q or properties P and Q stand in relation W? I would hold that this is not a counterexample, on the ground that there cannot be disjunctive relations. But this is to appeal to a view that some philosophers would reject.

A second, and more radical approach, involves replacing condition (2) by:

> (2**) Relation R is not analyzable in terms of other universals of *any* order.

This more radical approach, which I believe is preferable, does necessitate another change in the account. For suppose that there is a nomological relation R_1 holding between properties *P* and *Q*, in virtue of which it is a law that everything with property *P* has property *Q*, and a nomological relation R_2 holding between properties *Q* and *S*, in virtue of which it is a law that everything with property *Q* has property *S*. It will then be a law that everything with property *P* has property *S*, and this may be so simply in virtue of the relations R_1 and R_2 which hold between properties *P* and *Q*, and *Q* and *S*, respectively, and not because of any additional relation holding between properties *P* and *S*. Given the revised account of nomological relations, and hence of nomological statements, the generalization that everything with property *P* has property *S* could not be classified as nomological. But this consequence can be avoided by viewing the revised account as concerned only with basic or underived nomological statements, and then defining nomological statements as those entailed by the class of basic nomological statements.

3. Laws and Nomological Statements

The class of nomological statements characterized in the preceding section does not seem to coincide with the class of laws. Suppose it is a nomological truth that $(x)(Px \supset Qx)$. Any statement entailed by this must also be nomological, so it will be a nomological truth that $(x)[(Px \wedge Rx) \supset Qx]$, regardless of what property *R* is. Now it is certainly true that the latter statement will be nomologically necessary, and thus, in a broad sense of "law," it will express a law. Nevertheless it is important for some purposes—such as the analysis of causal statements and subjunctive conditionals—to define a subclass of nomological statements to which such statements will not belong. Consider, for example, the nomological statement that all salt, when in water, dissolves. If this is true, it will also be a nomological truth that all salt, when both in water and in the vicinity of a piece of gold, dissolves. But one does not want to say that the cause of a piece of salt's dissolving was that it was in water and in the vicinity of some gold. In the description of causes one wants to exclude irrelevant facts. Or consider the counterfactual: "If this piece of salt were in water and were not dissolving, it would not be in the vicinity of a piece of gold." If one says that all nomological statements support counterfactuals, and that it is a nomological truth that all salt when both in water and near gold dissolves, one will be forced to accept the preceding counterfactual, whereas it is clear that there is good reason not to accept it.

Intuitively, what one wants to do is to define a subclass of nomological statements in the broad sense, containing only those involving no irrelevant conditions. Nomological statements belonging to this subclass will be laws.[7] But how is this class to be defined? It cannot be identified with the class of underived nomological statements, since it is certainly possible for laws in the strict sense to be entailed by other, more comprehensive laws. The most plausible answer, I think, emerges if one rewrites nomological statements in full disjunctive normal form.[8] Thus, if this is done for the statement $(x)[(Px \wedge Rx) \supset Qx]$, one has:

$$(x)[(Px \wedge Rx \wedge Qx) \vee (Px \wedge -Rx \wedge Qx) \vee (-Px \wedge Rx \wedge Qx) \vee (-Px \wedge -Rx \wedge Qx) \vee (Px \wedge -Rx \wedge -Qx) \vee (-Px \wedge Rx \wedge -Qx) \vee (-Px \wedge -Rx \wedge -Qx)].$$

Of the seven disjuncts that compose the matrix of the statement rewritten in this way, one is of special interest: $Px \wedge -Rx \wedge -Qx$. For if it is a nomological truth that $(x)(Px \supset Qx)$, it is nomologically impossible for anything to satisfy that disjunct. It is this feature, I suggest, that distinguishes between nomological statements in general and laws in the narrow sense. If so, the following is a natural analysis of the concept of a law:

S expresses a law if and only if

(1) S is a nomological statement, and

(2) there is no nomological statement *T* such that, when *S* is rewritten in full disjunctive normal form, there is a disjunct *D* in the matrix such that *T* entails that there will be nothing that satisfies *D*.

This account is not, however, entirely satisfactory. Suppose that the statements $(x)(Px \supset Qx)$ and $(x)(Rx \supset Qx)$ both express laws. Since the statement $(x)[(Px \supset Qx) \wedge (Rx \supset Rx)]$ is logically equivalent to the former statement, it presumably expresses the same law. But when the last statement is written in full disjunctive normal form, it contains the disjunct $-Px \wedge -Qx \wedge Rx$, which cannot be satisfied by anything if it is a law that $(x)(Rx \supset Qx)$.

So some revision is necessary. The natural move is to distinguish between essential and inessential occurrences of terms in a statement: a term occurs essentially in a statement *S* if and only if there is no logically equivalent statement *S** which does not contain an occurrence of the same term. The account can thus be revised to read:

S expresses a law if and only if

(1) *S* is a nomological statement, and

(2) there are no nomological statements *S** and *T* such that *S** is logically equivalent to *S*, all constant terms in *S** occur essentially, and when *S** is rewritten in full disjunctive normal form, there is a disjunct *D* in the matrix such that *T* entails that nothing will satisfy *D*.

4. Objections

In this section I shall consider three objections to the approach advocated here. The first is that the account offered of the truth conditions of nomological statements is in some sense ad hoc and unilluminating. The second objection is that the analysis commits one to a very strong version of realism with respect to universals. The third is that the account offered places an unjustifiable restriction upon the class of nomological statements.

The basic thrust of the first objection is this. There is a serious problem about the truth conditions of laws. The solution offered here is that there are relations—referred to as nomological—which hold among universals, and which function as truth-makers for laws. How does this solution differ from simply saying that there are special facts—call them nomological—which are the facts which make laws true? How is the one approach any more illuminating than the other?

The answer is two-fold. First, to speak simply of nomological facts does nothing to *locate* those facts; that is, to specify the individuals that are the constituents of the facts in question. In contrast, the view advanced here does locate the relevant facts: they are facts about universals, rather than facts about particulars. And support was offered for this contention, viz., that otherwise no satisfactory analysis of the truth conditions of basic laws without positive instances is forthcoming. Secondly, the relevant facts were not merely located, but *specified,* since not only the individuals involved, but their relevant attributes, were described. It is true that the attributes had to be specified theoretically, and hence in a sense indirectly, but this is also the case when one is dealing with theoretical terms attributing properties to particulars. The Lewis-type account of the meaning of theoretical terms that was appealed to is just as applicable to terms that refer to relations among universals—including nomological relations—as it is to terms that refer to properties of, and relations among, particulars.

Still, the feeling that there is something unilluminating about the account may persist. How does one *determine,* after all, that there is, in any given case, a nomological relation holding among universals? In what sense have truth conditions for nomological statements really been supplied if it remains a mystery how one answers the epistemological question?

I think this is a legitimate issue, even though I do not accept the verificationist claim that a statement has factual meaning only if it is in principle verifiable. In the next section I will attempt to show that, given my account of the truth conditions of nomological statements, it is possible to have evidence that makes it reasonable to accept generalizations as nomological.

The second objection is that the analysis offered involves a very strong metaphysical commitment. It is not enough to reject nominalism. For one must, in the first place, also hold that there are higher order universals which are not reducible to properties of, and relations among, particulars. And secondly, although the account may not entail that Platonic realism—construed minimally as the claim that there are some uninstantiated universals—is true, it does entail that if the world had been slightly different, it *would* have been true.

The first part of this objection does not seem to have much force as an objection. For what reasons are there for holding that there are no irreducible, higher order relations? On the other hand, it does point to a source of uneasiness which many are likely to feel. Semantics is usually done in a way that is compatible with nominalism. Truth conditions of sentences are formulated in terms of particulars, sets of particulars, sets of sets of particulars, and so on. The choice, as a metalanguage for semantics, of a language containing terms whose referents are either universals, or else intensional entities, such as concepts, though rather favored by the later Carnap[9] and others, has not been generally accepted. Once irreducible higher order relations enter into the account, a shift to a metalanguage containing terms referring to entities other than particulars appears unavoidable, and I suspect that this shift in metalanguages may make the solution proposed here difficult for many to accept.

The second part of this objection raises a deeper, and more serious point. Does it follow from the analysis offered that, if the world had been somewhat different, Platonic realism would have been true? The relevant argument is this. Suppose that materialism is false, and that there is, for example, a nonphysical property of being an experience of the red variety. Then consider what our world would have been like if the earth had been slightly closer to the sun, and if conditions in other parts of the universe had been such that life evolved nowhere else. The universe would have contained no sentient organism, and hence no experiences of the red variety. But wouldn't it have been true in *that* world that if the earth had been a bit farther from the sun life would have evolved, and there would have been experiences of the red variety? If so, in virtue of what would this subjunctive conditional have been true? Surely an essential part of what would have made it true is the existence of a certain psychophysical law linking complex physical states to experiences of the red variety. But if the truthmakers for laws are relations among universals, it could not be a law in that world that whenever a complex physical system is in a certain state, there is an experience of the red variety, unless the property of being an experience of the red variety exists in that world. Thus, if the account of laws offered above is correct, one can describe a slightly altered version of our world in which there would be uninstantiated, and hence transcendent, universals.

The argument, as stated, is hardly conclusive. It does depend, for example, on the assumption that materialism is false. However this assumption is not really necessary. All that is required is the assumption that there are emergent properties. It makes no difference to the argument whether such emergent properties are physical or nonphysical.

If our world does not contain any emergent properties, it will not be possible to argue that if our world had been slightly different, there would have been uninstantiated universals. However one can argue for a different conclusion that may also seem disturbing. For if it is granted, not that there are emergent properties, but only that the concept of an emergent property is coherent, then it can be argued that the existence of uninstantiated, and hence transcendent, universals is logically possible. A conclusion that will commend itself to few philosophers who reject Platonic realism, since the arguments usually directed against Platonic realism, if sound, show that it is *necessarily* false.

Another way of attempting to avoid the conclusion is by holding that if the world had been different in the way indicated, there would have been no psychophysical laws. This view may be tenable, although it strikes me as no more plausible than the stronger contention that there cannot be basic laws that lack positive instances. As a result, I am inclined to accept the contention that if the account of laws set out above is correct, there is reason to believe that Platonic realism, construed only as the doctrine that there are uninstantiated universals, is not incoherent.

The final objection to be considered is that there are statements that it would be natural, in some possible worlds, to view as nomological, but which would not be so classified on the account given here. Suppose, for example, the world were as follows. All the fruit in Smith's garden at any time are apples. When one attempts to take an orange into the garden, it turns into an elephant. Bananas so treated become apples as they cross the boundary, while pears are resisted by a force that cannot be overcome. Cherry trees planted in the garden bear apples, or they bear nothing at all. If all these things were true, there would be a very strong case for its being a law that all the fruit in Smith's garden are apples. And this case would be in no way undermined if it were found that no other gardens, however similar to Smith's in all other respects, exhibited behavior of the sort just described.

Given the account of laws and nomological statements set out above, it cannot, in the world described, be a law, or even a nomological statement, that all the fruit in Smith's garden are apples. If relations among universals are the truth-makers for nomological statements, a statement that contains essential reference to a specific particular can be nomological only if entailed by a corresponding, universally quantified statement free of such reference. And since, by hypothesis, other gardens do not behave as Smith's does, such an entailment does not exist in the case in question.

What view, then, is one to take of the generalization about the fruit in Smith's garden, in the world envisaged? One approach is to say that although it cannot be a law, in that world, that all the fruit in Smith's garden are apples, it can be the case that there is some property *P* such that Smith's garden has property *P*, and it is a law that all the fruit in any garden with property *P* are apples. So that even though it is not a nomological truth that all the fruit in Smith's garden are apples, one can, in a loose sense, speak of it as "derived" from a nomological statement.

This would certainly seem the most natural way of regarding the generalization about Smith's garden. The critical question, though, is whether it would be reasonable to maintain this view in the face of any conceivable evidence. Suppose that careful investigation, over thousands of years, has not uncovered any difference in intrinsic properties between Smith's garden and other gardens, and that no experimental attempt to produce a garden that will behave as Smith's does has been successful. Would it still be reasonable to postulate a theoretical property *P* such that it is a law that all the fruit in gardens with property *P* are apples? This issue strikes me as far from clear, but I incline slightly to a negative answer. For it would seem that, given repeated failures to produce gardens that behave as Smith's garden does, one might well be justified in concluding that if there is such a property *P*, it is one whose exemplification outside of Smith's garden is nomologically impossible. And this seems like a strange sort of property to be postulating.

I am inclined to think, then, that it is logically possible for there to be laws and nomological statements, in the strict sense, that involve ineliminable reference to specific individuals. But it does not matter, with regard to the general view of laws advanced here, whether that is so. If the notion of nomological statements involving ineliminable reference to specific individuals turns out to be conceptually incoherent, the present objection will be mistaken. Whereas if there can be such nomological statements, my account requires only minor revision to accommodate them. The definition of a construction function will have to be changed so that instances of universals, and not merely universals, can be elements of the ordered *n*-tuples that it takes as arguments, and the definition of a nomological relation will have to be similarly altered, so that it can be a relation among both universals and instances of universals. These alterations will result in an analysis of the truth conditions of nomological statements that allows for the possibility of ineliminable reference to specific particulars. And they will do so, moreover, without opening the door to accidentally true generalizations. Both condition (2*) and condition (2**), as set out in section 2 above, appear sufficiently strong to block such counterexamples.

5. The Epistemological Question

In the previous section I mentioned, but did not discuss, the contention that the analysis of nomological statements advanced here is unilluminating because it does not provide any account of how one determines whether a given nomological relation holds among specific universals. I shall argue that this objection is mistaken, and that in fact one of the merits of the present view is that it does make possible an answer to the epistemological question.

Suppose, then, that a statement is nomological if there is an irreducible theoretical relation holding among certain universals which necessitates the statement's being true. If this is right, there should not be any special epistemological problem about the grounds for accepting a given statement as nomological. To assert that a statement is nomological will be to advance a theory claiming that certain universals stand in some nomological relation. So one would think that whatever account is to be offered of the grounds for accepting theories as true should also be applicable to the case of nomological statements.

I should like, however, to make plausible, in a more direct fashion, this claim that the sorts of considerations that guide our choices among theories also provide adequate grounds for preferring some hypotheses about nomological statements to others. Consider, then, a familiar sort of example. John buys a pair of pants, and is careful to put only silver coins in the right hand pocket. This he does for several years, and then destroys the pants. What is the status of the generalization: "Every coin in the right hand pocket of John's pants is silver"? Even on the limited evidence described, one would not be justified in accepting it as nomological. One of the central types of considerations in support of a theory is that it in some sense provides the best explanation of certain observed states. And while the hypothesis that there is a theoretical relation *R* which holds among the universals involved in the proposition that every coin in the right hand pocket of John's pants is silver, and which necessitates that proposition's being true, does explain why there were only silver coins in the pocket, this explanation is unnecessary. Another explanation is available, namely, that John wanted to put only silver coins in that pocket, and carefully inspected every coin to ensure that it was silver before putting it in.

If the evidence is expanded in certain ways, the grounds for rejecting the hypothesis that the statement is nomological become even stronger. Suppose that one has made a number of tests on other pants, ostensibly similar to John's, and found that the right hand pockets accept copper coins as readily as silver ones. In the light of this additional evidence, there are two main hypotheses to be considered:

> H_1: It is nomologically possible for the right hand pocket of any pair of pants of type *T* to contain a nonsilver coin;
>
> H_2: There is a pair of pants of type *T*, namely John's, such that it is nomologically impossible for the right hand pocket to contain a nonsilver coin; however all other pairs of pants of type *T* are such that it is nomologically possible for the right hand pocket to contain a nonsilver coin.

H_1 and H_2 are conflicting hypotheses, each compatible with all the evidence. But it is clear that H_1 is to be preferred to H_2. First, because H_1 is simpler than H_2, and secondly, because the generalization explained by H_2 is one that we already have an explanation for.

Let us now try to get clearer, however, about the sort of evidence that provides the strongest support for the hypothesis that a given generalization is nomological. I think the best way of doing this is to consider a single generalization, and to ask, first, what a world would be like in which one would feel that the generalization was merely accidentally true, and then, what changes in the world might tempt one to say that the generalization was not accidental, but nomological. I have already sketched a case of this sort. In our world, if all the fruit in Smith's garden are apples, it is only an accidentally true generalization. But if the world were different in certain ways, one might classify the generalization as nomological. If one never succeeded in getting pears into Smith's garden, if bananas changed into apples, and oranges into elephants as they crossed the boundary, etc., one might well be tempted to view the generalization as a law. What we now need to do is to characterize the evidence that seems to make a critical difference. What is it about the sort of events described that makes them significant? The answer, I suggest, is that they are events which determine which of "conflicting" generalizations are true. Imagine that one has just encountered Smith's garden. There are many generalizations that one accepts—generalizations that are supported by many positive instances, and for which no counterexamples are known, such as "Pears thrown with sufficient force towards nearby gardens wind up inside them," "Bananas never disappear, nor change into other things such as apples," "Cherry trees bear only cherries as fruit." One notices that there are many apples in the garden, and no other fruit, so the generalization that all the fruit in Smith's garden are apples is also supported by positive instances, and is without counterexamples. Suppose now that a banana is moving in the direction of Smith's garden. A partial conflict situation exists, in that there are some events which will, if they occur, falsify the generalization that all the fruit in Smith's garden are apples, and other events which will falsify the generalization that bananas never change into other objects. There are, of course, other possible events which will falsify neither generalization: the banana may simply stop moving as it reaches the boundary. However there may well be other generalizations that one accepts which will make the situation one of inevitable conflict, so that whatever the outcome, at least one generalization will be falsified. Situations of inevitable conflict can arise even for two generalizations, if related in the proper way. Thus, given the generalizations that $(x)(Px \supset Rx)$ and that $(x)(Qx \supset -Rx)$, discovery of an object b such that both Pb and Qb would be a situation of inevitable conflict.

Many philosophers have felt that a generalization's surviving such situations of conflict, or potential falsification, provides strong evidence in support of the hypothesis that the generalization is nomological. The problem, however, is to *justify* this view. One of the merits of the account of the nature of laws offered here is that it provides such a justification.

The justification runs as follows. Suppose that one's total evidence contains a number of supporting instances of the generalization that $(x)(Px \supset Rx)$, and of the generalization that $(x)(Qx \supset -Rx)$, and no evidence against either. Even such meager evidence may provide some support for the hypothesis that these generalizations are nomological, since the situation may be such that the only available explanation for the absence of counterexamples to the generalizations is that there are theoretical relations holding among universals which necessitate those generalizations. Suppose now that a conflict situation arises: an object *b* is discovered which is both *P* and *Q*. This new piece of evidence will reduce somewhat the likelihood of both hypotheses, since it shows that at least one of them must be false. Still, the total evidence now available surely lends some support to both hypotheses. Let us assume that it is possible to make at least a rough estimate of that support. Let *m* be the probability, given the available evidence, that the generalization $(x)(Px \supset Rx)$ is nomological, and *n* the probability that the generalization $(x)(Qx \supset -Rx)$ is nomological—where $(m + n)$ must be less than or equal to one. Suppose finally that *b*, which has property *P* and property *Q*, turns out to have property *R*, thus falsifying the second generalization. What we are now interested in is the effect this has upon the probability that the first generalization is nomological. This can be calculated by means of Bayes' Theorem:

> Let *S* be: It is a nomological truth that $(x)(Px \supset Rx)$.
>
> Let *T* be: It is a nomological truth that $(x)(Qx \supset -Rx)$.
>
> Let H_1 be: *S* and not-*T*.
>
> Let H_2 be: Not-*S* and *T*.
>
> Let H_3 be: *S* and *T*.
>
> Let H_4 be: Not-*S* and not-*T*.
>
> Let *E* describe the total antecedent evidence, including the fact that *Pb* and *Qb*

Let E^* be: E and Rb.

Then Bayes' Theorem states:

Probability (H_1, given that E* and E) =

$$\frac{\text{Probability } (E^*, \text{given that } H_1 \text{ and } E) \times \text{Probability } (H_1 \text{ and } E)}{\sum_{i=1}^{4} [\text{Probability } (E^*, \text{given that } H_i \text{ and } E) \times \text{Probability } (H_1 \text{ and } E)]}$$

Taking the antecedent evidence as given, we can set the probability of E equal to one. This implies that the probability of H_i and E will be equal to the probability of H_i given that E.

Probability (H_3, given that E)

= Probability (S and T, given that E)

= O, since E entails that either not-S or not-T.

Probability (H_1, given that E)

= Probability (S and not-T, given that E)

= Probability (S, given that E)

Probability (S and T, given that E)

= $m - O = m$.

Similarly, probability (H_2, given that E) = n.

Probability (H_4, given that E)

= 1 - [Probability (H_1, given that E) + Probability (H_2, given that E) + Probability (H_3, given that E)]

= $1 - (m + n)$

Probability (E^*, given that H_i and E)

= 1, if $i = 1$ or $i = 3$

= 0, if $i = 2$

= some value k, if $i = 4$.

Bayes' Theorem then gives:

Probability (H_1, given that E^* and E)

$$= \frac{1 \times m}{(1 \times m) + (0 \times n) + (1 \times 0) + (k \times (1 - (m - n)))}$$

$$= \frac{m}{m + k(1 - m - n)}$$

In view of the fact that, given that E^*, it is not possible that T, this value is also the probability that S, that is, the likelihood that the generalization that $(x)(Px \supset Rx)$ is nomological. Let us consider some of the properties of this result. First, it is easily seen that, provided neither m nor n is equal to zero, the likelihood that the generalization is nomological will be greater after its survival of the conflict situation than it was before. The value of $m/[m + k(1 - m - n)]$ will be smallest when k is largest. Setting k equal to one gives the value $m/[m + 1 - m - n]$ i.e., $m/[1-n]$, and this is greater than m if neither m nor n is equal to zero.

Secondly, the value of $m/[m + k(1 - m - n)]$ increases as the value of k decreases, and this too is desirable. If the event that falsified the one generalization were one that would have been very likely if neither generalization had been nomological, one would not expect it to lend as much support to the surviving generalization as it would if it were an antecedently improbable event.

Thirdly, the value of $m/[m + k(1 - m - n)]$ increases as n increases. This means that survival of a conflict with a well-supported generalization results in a greater increase in the likelihood that a generalization is nomological than survival of a conflict with a less well-supported generalization. This is also an intuitively desirable result.

Finally, it can be seen that the evidence provided by survival of conflict situations can quickly raise the likelihood that a generalization is nomological to quite high values. Suppose, for example, that $m = n$, and that $k = 0.5$. Then $m/[m + k(1 - m - n)]$ will be equal to $2m$. This result agrees with the view that laws, rather than being difficult or impossible to confirm, can acquire a high degree of confirmation on the basis of relatively few observations, provided that those observations are of the right sort.

But how is this justification related to the account I have advanced as to the truth conditions of nomological statements? The answer is that there is a crucial assumption that seems reasonable if relations among universals are the truth-makers for laws, but not if facts about particulars are the truth-makers. This is the assumption that m and n are not equal to zero. If one takes the view that it is facts about the particulars falling under a generalization that make it a law, then, if one is dealing with an infinite universe, it is hard to see how one can be justified in assigning any non-zero probability to a generalization, given evidence concerning only a finite number of instances. For surely there is some non-zero probability that any given particular will falsify the generalization, and this entails, given standard assumptions, that as the number of particulars becomes infinite, the probability that the generalization will be true is, in the limit, equal to zero.[10]

In contrast, if relations among universals are the truth-makers for laws, the truth-maker for a given law is, in a sense, an "atomic" fact, and it would seem perfectly justified, given standard principles of confirmation theory, to assign some non-zero probability to this fact's obtaining. So not only is there an answer to the epistemological question; it is one that is only available given the type of account of the truth conditions of laws advocated here.

6. Advantages of This Account of Nomological Statements

Let me conclude by mentioning some attractive features of the general approach set out here. First, it answers the challenge advanced by Chisholm in his article "Law Statements and Counterfactual Inference":[11]

> Can the relevant difference between law and non-law statements be described in familiar terminology without reference to counterfactuals, without use of modal terms such as "causal necessity," "necessary condition," "physical possibility," and the like, and without use of metaphysical terms such as "real connections between matters of fact"?

The account offered does precisely this. There is no reference to counterfactuals. The notions of logical necessity and logical entailment are used, but no nomological modal terms are employed. Nor are there any metaphysical notions, unless a notion such as a contingent relation among universals is to be counted as metaphysical. The analysis given involves nothing beyond the concepts of logical entailment, irreducible higher order universals, propositions, and functions from ordered sets of universals into propositions.

A second advantage of the account is that it contains no reference to possible worlds. What makes it true that statements are nomological are not facts about dubious entities called possible worlds, but facts about the actual world. True, these facts are facts about universals, not about particulars, and they are theoretical facts, not observable ones. But neither of these things should worry one unless one is either a reductionist, at least with regard to higher order universals, or a rather strict verificationist.

Thirdly, the account provides a clear and straightforward answer to the question of the difference between nomological statements and accidentally true generalizations: a generalization is accidentally true in virtue of facts about particulars; it is a nomological truth in virtue of a relation among universals.

Fourthly, this view of the truth conditions of nomological statements explains the relationships between different types of generalizations and counterfactuals. Suppose it is a law that $(x)(Px \supset Qx)$. This will be so in virtue of a certain irreducible relation between the universals P and Q. If now one asks what would be the case if some object b which at present lacks property P were to have P, the answer will be that this supposition about the particular b does not give one any reason for supposing that the universals P and Q no longer stand in a relation of nomic necessitation, so one can conjoin the supposition that b has property P with the proposition that the nomological relation in question holds between P and Q, from which it will follow that b has property Q. And this is why one is justified in asserting the counterfactual "If b were to have property P, it would also have property Q."

Suppose instead that it is only an accidentally true generalization that $(x)(Px \supset Qx)$. Here it is facts about particulars that make the generalization true. So if one asks what would be the case if some particular b which lacks property P were to have P, the situation is very different. Now one is supposing an alteration in facts that may be relevant to the truth conditions of the generalization. So if object b lacks property Q, the appropriate conclusion may be that if b were to have property P, the generalization that (x)

($Px \supset Qx$) would no longer be true, and thus that one would not be justified in conjoining that generalization with the supposition that *b* has property *P* in order to support the conclusion that *b* would, in those circumstances, have property *Q*. And this is why accidentally true generalizations, unlike laws, do not support the corresponding counterfactuals.

Fifthly, this account of nomological statements allows for the possibility of even basic laws that lack positive instances. And this accords well with our intuitions about what laws there would be in cases such as a slightly altered version of our own world, in which life never evolves, and in that of the universe with the two types of fundamental particles that never meet.

Sixthly, it is a consequence of the account given that if *S* and *T* are logically equivalent sentences, they must express the same law, since there cannot be a nomological relation among universals that would make the one true without making the other true. I believe that this is a desirable consequence. However some philosophers have contended that logically equivalent sentences do not always express the same law. Rescher, for example, in his book *Hypothetical Reasoning,* claims that the statement that it is a law that all *X*'s are *Y*'s makes a different assertion from the statement that it is a law that all non-*Y*'s are non-*X*'s, on the grounds that the former asserts "All *X*'s are *Y*'s *and further* if *z* (which isn't an *X*) were an *X*, then *z* would be a *Y*," while the latter asserts "All non-*Y*'s are non-*X*'s *and further* if *z* (which isn't a non-*Y*) were a non-*Y*, then *z* would be a non-*X*."[12] But it would seem that the answer to this is simply that the statement that it is a law that all *X*'s are *Y*'s *also* entails that if *z* (which isn't a non-*Y*) were a non-*Y*, then *z* would be a non-*X*. So Rescher has not given us any reason for supposing that logically equivalent sentences can express different laws.

The view that sentences which would normally be taken as logically equivalent may, when used to express laws, not be equivalent, has also been advanced by Stalnaker and Thomason. Their argument is this. First, laws can be viewed as generalized subjunctive conditionals. "All *P*'s are *Q*'s," when stating a law, can be analyzed as "For all *x*, if *x* were a *P* then *x* would be a *Q*." Secondly, contraposition does not hold for subjunctive conditionals. It may be true that if *a* were *P* (at time t_1) then *a* would be *Q* (at time t_2) yet false that if *a* were not *Q* (at time t_2), then *a* would not have been *P* (at time t_1). Whence it follows that its being a law that all *P*'s are *Q*'s is not equivalent to its being a law that all non-*Q*'s are non-*P*'s.[13]

The flaw in this argument lies in the assumption that laws can be analyzed as generalized subjunctive conditionals. The untenability of the latter claim can be seen by considering any possible world *W* which satisfies the following conditions:

(1) The only elementary properties in *W* are *P*, *Q*, *F*, and *G*;

(2) There is some time when at least one individual in *W* has properties *F* and *P*, and some time when at least one individual in *W* has properties *F* and *Q*;

(3) It is true, but not a law, that everything has either property *P* or property *Q*;

(4) It is a law that for any time *t*, anything possessing properties *F* and *P* at time *t* will come to have property *G* at a slightly later time *t**;

(5) It is a law that for any time *t*, anything possessing properties *F* and *Q* at time *t* will come to have property *G* at a slightly later time *t**;

(6) No laws are true in *W* beyond those entailed by the previous two laws.

Consider now the generalized subjunctive conditional, "For all *x*, and for any time *t*, if *x* were to have property *F* at time *t*, *x* would come to have property *G* at a slightly later time *t**." This is surely true in *W*, on any plausible account of the truth conditions of subjunctive conditionals. For let *x* be any individual in *W* at any time *t*. If *x* has *P* at time *t*, then in view of the law referred to at (4), it will be true that if *x* were to have *F* at *t*, it would come to have *G* at *t**. While if *x* has *Q* at *t*, the conditional will be true in virtue of that fact together with the law referred to at (5). But given (3), *x* will, at time *t*, have either property *P* or property *Q*. So it will be true in *W*, for any *x* whatsoever, that if *x* were to have property *F* at time *t*, it would come to have property *G* at a slightly later time *t**.

If now it were true that laws are equivalent to generalized subjunctive conditionals, it would follow that it is a law in *W* that for every *x*, and every time *t*, if *x* has *F* at *t*, then *x* will come to have *G* at *t**. But this law does not follow from the laws referred to at (4) and (5), and hence is excluded by condition (6). The possibility of worlds such as *W* shows that laws are not equivalent to generalized subjunctive conditionals. As a result, the Stalnaker-Thomason argument is unsound, and there is no reason for thinking that logically equivalent statements can express non-equivalent laws.

Seventhly, given the above account of laws and nomological statements, it is easy to show that such statements have the logical properties one would naturally attribute to them. Contraposition holds for laws and nomological statements, in view of the fact that logically equivalent statements express the same law. Transitivity also holds: if it is a law (or nomological statement) that $(x)(Px \supset Qx)$, and that $(x)(Qx \supset Rx)$, then it is also a law (or nomological statement) that $(x)(Px \supset Rx)$. Moreover, if it is a law (or nomological statement) that $(x)(Px \supset Qx)$, and that $(x)(Px \supset Rx)$, then it is a law (or nomological statement) that $(x)[Px \supset (Qx \wedge Rx)]$, and conversely. Also, if it is a law (or nomological statement) that $(x)\,[(Px \supset Qx) \supset Rx]$, then it is a law (or nomological statement) both that $(x)(Px \supset Rx)$, and that $(x)(Qx \supset Rx)$, and conversely. And in general, I think that laws and nomological statements can be shown, on the basis of the analysis proposed here, to have all the formal properties they are commonly thought to have.

Eighthly, the account offered provides a straightforward explanation of the nonextensionality of nomological contexts. The reason that it can be a law that $(x)(Px \supset Rx)$, and yet not a law that $(x)(Qx \supset Rx)$, even if it is true that $(x)(Px \equiv Qx)$, is that, in view of the fact that the truth-makers for laws are relations among universals, the referent of the predicate *"P"* in the sentence "it is a law that $(x)(Px \supset Rx)$" is, at least in the simplest case, a universal, rather than the set of particulars falling under the predicate. As a result, interchange of co-extensive predicates in nomological contexts may alter the referent of part of the sentence, and with it, the truth of the whole.

Finally, various epistemological issues can be resolved given this account of the truth conditions of nomological statements. How can one establish that a generalization is a law, rather than merely accidentally true? The general answer is that if laws hold in virtue of theoretical relations among universals, then whatever account is to be given of the grounds for accepting theories as true will also be applicable to laws. The latter will not pose any independent problems. Why is it that the results of a few carefully designed experiments can apparently provide very strong support for a law? The answer is that if the truth-makers for laws are relations among universals, rather than facts about particulars, the assignment of non-zero initial probability to a law ceases to be unreasonable, and one can then employ standard theorems of probability theory, such as Bayes' Theorem, to show how a few observations of the right sort will result in a probability assignment that quickly takes on quite high values.

To sum up, the view that the truth-makers for laws are irreducible relations among universals appears to have much to recommend it. For it provides not only a noncircular account of the truth conditions of nomological statements, but an explanation of the formal properties of such statements, and a solution to the epistemological problem for laws.

Notes

I am indebted to a number of people, especially David Armstrong, David Bennett, Mendel Cohen, Michael Dunn, Richard Routley, and the editors of the *Canadian Journal of Philosophy* for helpful comments on earlier versions of this paper.

1. A vacuously true generalization is often characterized as a conditional statement whose antecedent is not satisfied by anything. This formulation is not entirely satisfactory, since it follows that there can be two logically equivalent generalizations, only one of which is vacuously true. A sound account would construe being vacuously true as a property of the content of a generalization, rather than as a property of the form of the sentence expressing the generalization.

2. See, for example, the incisive discussions by Jonathan Bennett in his article, "Counterfactuals and Possible Worlds," *Canadian Journal of Philosophy* 4 (1974), pages 381–402, and, more recently, by Frank Jackson in his article, "A Causal Theory of Counterfactuals," *Australasian Journal of Philosophy* 55 (1977), pages 3–21.

3. F. P. Ramsey, "General Propositions and Causality," in *The Foundations of Mathematics,* edited by R. B. Braithwaite, Paterson, New Jersey, 1960, page 242. The view described in the passage is one which Ramsey had previously held rather than the view he was setting out in the paper itself. For a sympathetic discussion of Ramsey's earlier position, see pages 72–77 of David Lewis's *Counterfactuals,* Cambridge, Massachusetts, 1973.

4. *Journal of Philosophy* 67 (1970), pages 427–46.

5. This problem was pointed out by Israel Scheffler in his book, *The Anatomy of Inquiry,* New York, 1963. See section 21 of part II, pages 218ff.

6. This version of the general theory is essentially that set out by David Armstrong in his 1978 book, *Universals and Scientific Realism.* In revising the present paper I have profited from discussions with Armstrong about the general theory, and the merits of our competing versions of it.

7. This interpretation of the expressions "nomological statement" and "law" follows that of Hans Reichenbach in his book, *Nomological Statements and Admissible Operations,* Amsterdam, 1954. It may be that the term "law" is ordinarily used in a less restricted sense. However there is an important distinction to be drawn here, and it seems natural to use the term "law" in perhaps a slightly narrower sense in order to have convenient labels for these two classes of statements.

8. This method of handling the problem was employed by Hans Reichenbach in *Nomological Statements and Admissible Operations.*

9. Compare Carnap's discussion in the section entitled "Language, Modal Logic, and Semantics," especially pages 889–905, in *The Philosophy of Rudolf Carnap,* edited by Paul A. Schilpp, La Salle, Illinois, 1963.

10. See, for example, Rudolf Carnap's discussion of the problem of the confirmation of universally quantified statements in section F of the appendix of his book, *The Logical Foundations of Probability,* 2nd edition, Chicago, 1962, pages 570–71.

11. Roderick M. Chisholm, "Law Statements and Counterfactual Inference," *Analysis* 15 (1955), pages 97–105, reprinted in *Causation and Conditionals,* edited by Ernest Sosa, London, 1975. See page 149.

12. Nicholas Rescher, *Hypothetical Reasoning,* Amsterdam, 1964, page 81.

13. Robert C. Stalnaker and Richmond H. Thomason, "A Semantical Analysis of Conditional Logic," *Theoria* 36 (1970), pages 39–40.

3

Do the Laws of Physics State the Facts?

NANCY CARTWRIGHT

0. Introduction. There is a view about laws of nature that is so deeply entrenched that it doesn't even have a name of its own. It is the view that laws of nature describe facts about reality. If we think that the facts described by a law obtain, or at least that the facts which obtain are sufficiently like those described in the law, we count the law true, or true-for-the-nonce, until further facts are discovered. I propose to call this doctrine the *facticity* view of laws. (The name is due to John Perry.)

It is customary to take the fundamental explanatory laws of physics as the ideal. Maxwell's equations, or Schroedinger's, or the equations of general relativity are paradigms, paradigms upon which all other laws—laws of chemistry, biology, thermodynamics, or particle physics—are to be modeled. But this assumption confutes the facticity view of laws. For the fundamental laws of physics do not describe true facts about reality. Rendered as descriptions of facts, they are false; amended to be true, they lose their fundamental, explanatory force.

To understand this claim, it will help to contrast biology with physics. J. J. C. Smart ([10]: chapter 2) has argued that biology is a second-rate science. This is because biology has no genuine laws of its own. It resembles engineering. Any general claim about a complex system, such as a radio or a living organism, will be likely to have exceptions. The generalizations of biology, or engineering's rules of thumb, are not true laws because they are not exceptionless. If this is a good reason, then it must be physics which is the second-rate science. Not only do the laws of physics have exceptions; unlike biological laws, they are not even true for the most part, or approximately true.

From *Pacific Philosophical Quarterly* 61 (1980): 75–84. Reprinted by permission of Blackwell Publishing.

The view of laws with which I begin—"Laws of nature describe facts about reality"—is a pedestrian view that, I imagine, any scientific realist will hold. It supposes that laws of nature tell how objects of various kinds behave: how they behave some of the time, or all of the time, or even (if we want to prefix a necessity operator) how they must behave. What is critical is that they talk about objects—real concrete things that exist here in our material world, things like quarks, or mice, or genes; and they tell us what these objects do.

Biological laws provide good examples. For instance, here is a generalization taken from a Stanford text on chordates (Alexander [1]: 179):

> The gymnotoids [American knife fish] are slender fish with enormously long anal fins, which suggest the blade of a knife of which the head is a handle. They often swim slowly with the body straight by undulating this fin. They [presumably "always" or "for the most part"] are found in Central and South America . . . Unlike the characins they ["usually"?] hide by day under river banks or among roots, or even bury themselves in sand, emerging only at night.

The fundamental laws of physics, by contrast, do not tell what the objects in their domain do. If we try to think of them in this way, they are simply false, not only false but deemed false by the very theory which maintains them. But if physics' basic, explanatory laws do not describe how things behave, what do they do? Once we have given up facticity, I don't know what to say. Richard Feynman, in *The Character of Physical Law,* offers an idea, a metaphor. Feynman tells us "There is . . . a rhythm and a pattern between the phenomena of nature which is not apparent to the eye, but only to the eye of analysis; and it is these rhythms and patterns which we call Physical Laws" ([3]: 13). Most philosophers will want to know a lot more about how these rhythms and patterns function. But at least Feynman does not claim that the laws he studies describe the facts.

I say that the laws of physics do not provide true descriptions of reality. This sounds like an anti-realist doctrine. Indeed it is, but to describe the claim in this way may be misleading. For anti-realist views in the philosophy of science are traditionally of two kinds. Bas van Fraassen [4] is a modern advocate of one of these versions of anti-realism; Hilary Putnam ([8] [9]) of the other. Van Fraassen is a sophisticated instrumentalist. He worries about the existence of unobservable entities, or rather, about the soundness of our grounds for believing in them; and he worries about the evidence which is supposed to support our theoretical claims about how these entities behave. But I have no quarrel with theoretical entities; and for the moment I am not concerned with how we know what they do. What is troubling me here is that our explanatory laws don't tell us what they do. It is in fact part of their explanatory role not to tell.

Hilary Putnam in his new version of transcendental realism also maintains that the laws of physics don't represent facts about reality. But this is because nothing—not even the most commonplace claim about the cookies which are burning in the oven—represents facts about reality. If anything did, Putnam would probably think that the basic equations of modern physics did best. This is the claim that I reject. I think we can allow that all sorts of statements represent facts of nature, including the generalizations one learns in biology or engineering. It is just the fundamental explanatory laws that don't truly represent. Putnam is worried about meaning and reference and how we are trapped in the circle of words. I am worried about truth and explanation, and how one excludes the other.

I. *Explanation by composition of causes, and the trade-off of truth and explanatory power.* Let me begin with a law of physics everyone knows—the law of universal gravitation. This is the law which Feynman uses for illustration; he endorses the view that this law is "the greatest generalization achieved by the human mind" (Feynman [3]: 14).

Law of Gravitation: $F = Gmm'/r^2$.

In words, Feynman tells us ([3]: 14),

> The Law of Gravitation is that two bodies exert a force between each other which varies inversely as the square of the distance between them, and varies directly as the product of their masses.

Does this law truly describe how bodies behave?

Assuredly not. Feynman himself gives one reason why. "Electricity also exerts forces inversely as the square of the distance, this time between charges" ([3]: 30). It's not true that for *any* two bodies the force between them is given by the law of gravitation. Some bodies are charged bodies, and the force between them is not Gmm'/r^2. Rather it is some resultant of this force with the electric force which Feynman refers to.

For bodies which are both massive and charged, the law of universal gravitation and Coulomb's law (the law which gives the force between two charges) interact to determine the final force. But neither law by itself truly describes how the bodies behave. No charged objects will behave just as the law of universal gravitation says; and any massive objects will constitute a counterexample to Coulomb's law. These two laws are not true; worse, they are not even approximately true. In the interaction between the electrons and the protons of an atom, for example, the Coulomb effect swamps the gravitational one, and the force which actually occurs is very different from that described by the law of gravity.

There is an obvious rejoinder: I have not given a complete statement of these two laws, only a shorthand version. The Feynman version has an implicit *ceteris paribus* modifier in front, which I have suppressed. Speaking more carefully, the law of universal gravitational is something like this:

> *If* there are no forces other than gravitational forces at work, *then* two bodies exert a force between each other which varies inversely as the square of the distance between them, and varies directly as the product of their masses.

I will allow that this law is a true law, or at least one which is held true within a given theory. But it is not a very useful law. One of the chief jobs of the law of gravity is to help explain the forces which objects experience in various complex circumstances. *This* law can explain in only very simple, or ideal circumstances. It can account for why the force is as it is when just gravity is at work; but it is of no help for cases in which both gravity and electricity matter. Once the *ceteris paribus* modifier has been attached, the law of gravity is irrelevant to the more complex and interesting situations.

This unhappy feature is characteristic of explanatory laws. I said that the fundamental laws of physics do not represent the facts, whereas biological laws and principles of engineering do. This statement is both too strong and too weak. Some laws of physics do represent facts, and some laws of biology—particularly the explanatory laws—do not. The failure of facticity does not have so much to do with the nature of physics, but rather with the nature of explanation. We think that nature is governed by a small number of simple, fundamental laws. The world is full of complex and varied phenomena, but these are not fundamental. They arise from the interplay of simpler processes obeying the basic laws of nature.

This picture of how nature operates to produce the subtle and complicated effects we see around us is reflected in the explanations that we give: we explain complex phenomena by reducing them to their simpler components. This is not the only kind of explanation we give, but it is an important and central kind. I shall use the language of John Stuart Mill, and call this *explanation by composition of causes* (Mill [7]: book 3, chapter 6).

It is characteristic of explanations by composition of causes that the laws they employ fail to satisfy the requirement of facticity. The force of these explanations comes from the presumption that the explanatory laws "act" in combination just as they would "act" separately. It is critical, then, that the laws cited have the same form, in or out of combination. But this is impossible if the laws are to describe the actual behavior of objects. The actual behavior is the resultant of simple laws in combination. The effect which occurs is not an effect dictated by any one of the laws separately. In order to be true in the composite case, the law must describe one effect (the effect which actually happens); but to be explanatory, it must describe another. There is a trade-off here between truth and explanatory power.

II. *How vector addition introduces causal powers.* Our example, where gravity and electricity mix, is an example of the composition of forces. We know that forces add vectorially. Doesn't vector addition provide a simple and obvious answer to my worries? When gravity and electricity are both at work, two forces are produced, one in accord with Coulomb's law, the other according to the law of universal gravitation. Each law is accurate. Both the gravitational and the electric force are produced as described; the two forces then add together, vectorially, to yield the total "resultant" force.

The vector addition story is, I admit, a nice one. But it is just a metaphor. We add forces (or the numbers that represent forces) when we do calculations. Nature does not "add" forces. For the "component" forces are not there, in any but a metaphorical sense, to be added; and the laws which say they are there must also be given a metaphorical reading. Let me explain in more detail.

The vector addition story supposes that Feynman has left something out in his version of the law of gravitation. The way he writes it, it sounds as if the law describes the *resultant* force exerted between two bodies, rather than a component force—the force which is *produced between the two bodies in virtue of their gravitational masses* (or, for short, the force *due to gravity*). A better way to state the law would be

> Two bodies produce a force between each other (the force due to gravity) which varies inversely as the square of the distance between them, and varies directly as the product of their masses.

Similarly, for Coulomb's law

> Two charged bodies produce a force between each other (the force due to electricity) which also varies inversely as the square of the distance between them, and varies directly as the product of their masses.

These laws, I claim, do not satisfy the facticity requirement. They appear, on the face of it, to describe what bodies do: in the one case, the two bodies produce a force of size Gmm'/r^2; in the other, they produce a force of size qq'/r^2. But this cannot literally be so. For the force of size Gmm'/r^2 and the force of size qq'/r^2 are not real, occurrent forces. In interaction, a single force occurs—the force we call the "resultant"—and this force is neither the force due to gravity nor the electric force. On the vector addition story, the gravitational and the electric force are both produced, yet neither exists.

Mill would deny this. He thinks that in cases of the composition of causes, each separate effect does exist—it exists as *part* of the resultant effect, just as the left half of a table exists as part of the whole table. Mill's paradigm for composition of causes is mechanics. He says:

> In this important class of cases of causation, one cause never, properly speaking, defeats or frustrates another; both have their full effect. If a body is propelled in two directions by two forces, one tending to drive it to the north, and the other to the east, it is caused to move in a given time exactly as far in *both* directions as the two forces would separately have carried it. (Mill [7]: book 3, chapter 6)

Mill's claim is unlikely. Events may have temporal parts, but not parts of the kind Mill describes. When a body has moved along a path due northeast, it has traveled neither due north nor due east. The first half of the motion can be a part of the total motion; but no pure north motion can be a part of a motion which always heads northeast. (We learn this from Judy Jarvis Thomson's *Acts and Other Events.*) The lesson is even clearer if the example is changed a little: a body is pulled equally in opposite directions. It doesn't budge an inch, but on Mill's picture it has been caused to move both several feet to the left and several feet to the right. I realize, however, that intuitions are strongly divided on these cases, so in the next section I will present an example for which there is no possibility for seeing the separate effects of the composed causes as part of the effect which actually occurs.

It is implausible to take the force due to gravity and the force due to electricity literally as parts of the actually occurring force. Is there no way, then, to make sense of the story about vector addition? I think there is, but it involves giving up the facticity view of laws. We can preserve the truth of Coulomb's law and the law of gravitation by making them about something other than the facts—the laws can describe the causal powers that bodies have.

Hume taught that "the distinction, which we often make betwixt *power* and the *exercise* of it, is . . . without foundation" (Hume [5]: part 3, section 14). It is just Hume's illicit distinction that we need here: the law of gravitation claims that two bodies have the *power* to produce a force of size Gmm'/r^2. But they don't always succeed in the *exercise* of it. What they actually produce depends on what other powers are at work, and on what compromise is finally achieved among them. This may be the way we do sometimes imagine the composition of causes. But if so, the laws we use talk not about what bodies do, but about what powers they possess.

The introduction of causal powers will not be seen as a very productive starting point in our current era of moderate empiricism. Without doubt, we do sometimes think in terms of causal powers, so it would be foolish to maintain that the facticity view must be correct and the use of causal powers a total mistake. Still, facticity cannot be given up easily. We need an account of what laws are that connects them, on the one hand, with standard scientific methods for confirming laws, and on the other, with the use they are put to for prediction, construction, and explanation. If laws of nature are presumed to describe the facts, then there are familiar, detailed philosophic stories to be told about why a sample of facts is relevant to their confirmation, and how they help provide knowledge and understanding of what happens in nature. Any alternative account of what laws of nature do and what they say must serve at least as well; and no story I know of causal powers makes a very good start.

III. *A real example of the composition of causes.* The ground state of the carbon atom has five distinct energy levels (see diagram on page 78):

Physics texts commonly treat this phenomenon sequentially, in three stages. I shall follow the discussion of Albert Messiah in volume 2 of *Quantum Mechanics* [6]. In the first stage, the ground state energy is calculated by a central field approximation; and the single line (a) is derived. For some purposes, it is accurate to assume that only this level occurs. But some problems require a more accurate description. This can be provided by noticing that the central field approximation takes account only of the *average* value of the electrostatic repulsion of the inner shell electrons on the two outer

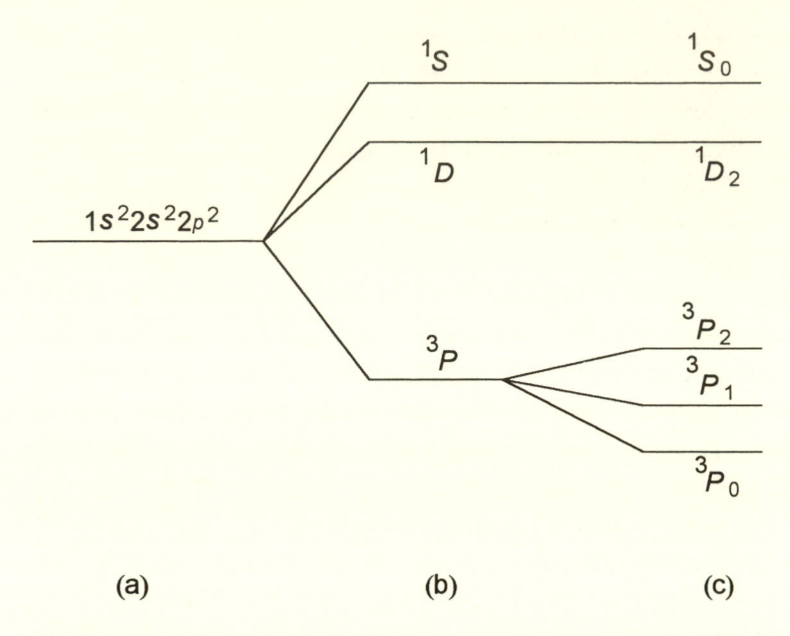

The levels of the ground state of the carbon atom; (a) in the central field approximation ($V_1 = V_2 = 0$); (b) neglecting spin-orbit coupling ($V_2 = 0$); (c) including spin-orbit coupling. (Messiah [6])

electrons. This defect is remedied at the second stage by considering the effects of a term which is equal to the difference between the exact Coulomb interaction and the average potential used in stage one. This corrective potential "splits" the single line (a) into three lines depicted in (b).

Still, the treatment is inaccurate because it neglects spin effects. Each electron has a spin, or internal angular momentum, and the spin of the electron couples with its orbital angular momentum to create an additional potential. The additional potential arises because the spinning electron has an intrinsic magnetic moment, and "an electron moving in [an electrostatic] potential 'sees' a magnetic field" ([6]: 552). About the results of this potential Messiah tells us, "Only the ^3P state is affected by the spin-orbit energy term; it gets split into three levels: $^3P_0$, $^3P_1$ and $^3P_2$" ([6]: 706). Hence the five levels pictured in (c).

The philosophic perplexities stand out most at the last stage. The five levels are due to a combination of a Coulomb potential, and a potential created by spin-orbit coupling "splits" the lowest of these again into three.

That is the explanation of the five levels. But how can we state the laws that it uses?

For the Coulomb effect we might try

> Whenever a Coulomb potential is like that in the carbon atom, the three energy levels pictured in (b) occur.

(The real law will of course replace "like that in the carbon atom" by a mathematical description of the Coulomb potential in carbon; and similarly for "the three energy levels pictured in (b).") The carbon atom itself provides a counterexample to this law. It has a Coulomb potential of the right kind; yet the five levels of (c) occur, not the three levels of (b).

We might, in analogy with the vector addition treatment of composite forces, try instead

> The energy levels produced by a Coulomb potential like that in the carbon atom are the three levels pictured in (b).

But (as with the forces "produced by gravity" in our earlier example) the levels which are supposed to be produced by the Coulomb potential are levels that don't occur. In actuality, five levels occur, and they do not include the three levels of (b). In particular, as we can see from Messiah's diagram, the lowest of the three levels—the ^3P—is not identical with any of the five. In the case of the composition of motions, Mill tried to see the "component" effects as parts of the actual effect. But that certainly will not work here. The ^3P level in (b) may be "split," and hence "give rise to," the $^3P_0$, $^3P_1$, and $^3P_2$ levels in (c); but it is certainly not a part of any of these levels.

It is hard to state a true factual claim about the effects of the Coulomb potential in the carbon atom. But quantum theory does guarantee that a certain *counterfactual* is true: the Coulomb potential, if it *were* the only potential at work, would produce the three levels in (b). Clearly this counterfactual bears on our explanation. But we have no model of explanation that shows how. The covering law model shows how statements of fact are relevant to explaining a phenomenon. But how is a truth about energy levels which would occur in quite different circumstances relevant to the levels which do occur in these? We think the counterfactual is important; but we have no account of how it works.

IV. *Composition of causes versus explanation by covering law.* The composition of causes is not the only method of explanation which can be employed. There are other methods, and some of these are compatible with the facticity view of laws. Standard covering law explanations are a prime example.

Sometimes these other kinds of explanation are available even when we give an explanation which tells what the component causes of a phenomenon are. For example, in the case of Coulomb's law and the law of gravity, we know how to write down a more complex law (a law with a more complex antecedent) which says exactly what happens when a system has both mass and charge. Mill thinks that such "super" laws are always available for mechanical phenomena. In fact, he thinks, "This explains why mechanics is a deductive or demonstrative science, and chemistry is not" (Mill [7]: book 3, chapter 6).

I want to make three remarks about these super laws and the covering explanations they provide: First, super laws aren't always available; second, even when they are available, they often don't explain much; third—and most importantly—even when other good explanations are to hand, if we fail to describe the component processes that go together to produce a phenomenon, we lose a central and important part of our understanding of what makes things happen.

(1) There are a good number of complex scientific phenomena which we are quite proud to be able to explain. For many of these explanations, however, super covering laws are not available to us. I argue this in "The Truth Doesn't Explain Much" (Cartwright [2]). Sometimes in these situations we have every reason to believe that a super law exists. (God has written it somewhere in the Book of Nature.) In other cases we have no good empirical reason to suppose even this much. (Nature may well be underdetermined; God failed to write laws for every complex situation.) Nevertheless, after we have seen what occurs in a specific case, we are often able to understand how various causes contributed to bring it about. We do explain, even without knowing the super laws. We need a philosophical account of laws and explanations which covers this very common scientific practice, and which shows why these explanations are good ones.

(2) Sometimes super laws, even when they are available to cover a case, may not be very explanatory. This is an old complaint against the covering law model of explanation: "Why does the quail in the garden bob its head up and down in that funny way whenever it walks?" . . . "Because they all do." In the example of spin-orbit coupling it does not explain the five ener-

gy levels that appear in a particular experiment to say "All carbon atoms have five energy levels."

(3) Often, of course, a covering law for the complex case will be explanatory. This is especially true when the antecedent of the law does not just piece together the particular circumstances that obtain on the occasion in question, but instead gives a more abstract description which fits with a general body of theory. In the case of spin-orbit coupling, Stephen Norman reminds that quantum mechanics provides general theorems about symmetry groups, and Hamiltonians, and degeneracies, from which we could expect to derive, covering law style, the energy levels of carbon from the appropriate abstract characterization of its Hamiltonian, and the symmetries it exhibits.

Indeed, we can do this; and if we don't do it, we will fail to see that the pattern of levels in carbon is a particular case of a general phenomenon which reflects a deep fact about the effects of symmetries in nature. On the other hand, to do only this misses the detailed causal story of *how* the splitting of spectral lines by the removal of symmetry manages to get worked out in each particular case.

This two-faced character is a widespread feature of explanation. Even if (contrary to what now, with the lack of success of the Wheeler-Misner program, seems to be a reasonable expectation) there is a single set of super laws which unifies all the complex phenomena one studies in physics (a geo-metro-electrodynamics, for example), our current picture may yet provide the ground for these laws: what the unified laws dictate to happen, happens *because of* the combined action of laws from separate domains, like the law of gravity and Coulomb's law. Without these laws, we would miss an essential portion of the explanatory story. Explanation by subsumption under super, unified covering laws would be no replacement for the composition of causes. It would be a complement. To understand how the consequences of the unified laws are brought about would require separate operation of the law of gravity, Coulomb's law, and so forth; and the failure of facticity for these contributory laws would still have to be faced.

V. *Conclusion.* There is a simple, straightforward view of laws of nature which is suggested by scientific realism, the facticity view: laws of nature describe how physical systems behave. This is by far the most common view, and a sensible one; but it doesn't work. It doesn't fit explanatory laws, like the fundamental laws of physics. Some other view is needed if we are to account for the use of laws in explanation; and I don't see any obvious candidate which is consistent with the realist's reasonable demand that laws describe reality and state facts which might well be true. There is, I have argued, a trade-off between factual content and explanatory power. We explain certain complex phenomena to be the result of the interplay of simple, fundamental laws. But what do these fundamental laws *say?* To play the role in explanation we demand of them, these laws must have the same form when they act together as when they act singly. In the simplest case, the consequences which the laws prescribe must be exactly the same in interaction, as the consequences which would obtain if the law were operating alone. But then, what the laws state cannot literally be true, for the consequences which would occur if it acted alone are not the consequences which actually occur when it acts in combination.

If we state the fundamental laws as laws about what happens when only a single cause is at work, then we can suppose the law to provide a true description. The problem arises when we try to take that law and use it to explain the very different things which happen when several causes are at work. This is the point of "The Truth Doesn't Explain Much" (Cartwright [2]). There is no difficulty writing down laws which we suppose to be true: "*If* there are no charges, no nuclear forces, . . . *then* the force between two masses of size m and m' *separated by a distance r is* Gmm'/r^2." We count this law true—what it says will happen, does happen—or at least happens to within a good approximation. But this law doesn't explain much. It is irrelevant to cases where there are electric or nuclear forces at work. The laws of physics, I concluded, to the extent that they are true, don't explain much. We could know all the true laws of nature, and still not know how to explain composite cases. Explanation must rely on something other than law.

But this view is absurd. There aren't two vehicles for explanation—laws for the rare occasions when causes occur separately; and another secret, nameless device for when they occur in combination. Explanations work in the same way whether one cause is at work, or many. "Truth Doesn't Explain" raises perplexities about explanation by composition of causes; and it concludes that explanation is a very peculiar scientific activity, which commonly does not make use of laws of nature. But scientific explanations do use laws. It is the laws themselves which are peculiar. The lesson to be learned is that the laws which explain by composition of causes fail to satisfy the facticity requirement. If the laws of physics are to explain how phenomena are brought about, they cannot state the facts.

Note

This paper was given as part of a symposium on "Explanation and Scientific Realism" at the School of Philosophy, University of Southern California, March 1980.

References

[1] Alexander, R. McNeill. *The Chordates* (Cambridge University Press, 1975).

[2] Cartwright, Nancy. "The Truth Doesn't Explain Much." Presented at the 54th Annual Meeting of the American Philosophical Association, Pacific Division, San Francisco, Calif., March 27–29, 1980. Preprinted in *American Philosophical Quarterly* 17 (1980).

[3] Feynman, Richard. *The Character of Physical Law* (Cambridge, Mass.: MIT Press, 1967).

[4] van Fraassen, Bas. *The Scientific Image* (Oxford University Press, 1980).

[5] Hume, David. *A Treatise of Human Nature* (edited by L.A. Selby Bigge. Oxford: Clarendon Press, 1978).

[6] Messiah, Albert. *Quantum Mechanics* (Amsterdam: North Holland, 1961).

[7] Mill, John Stuart. *A System of Logic* (London: John W. Parker and Son, 1856).

[8] Putnam, Hilary. *Meaning and the Moral Sciences* (London: Routledge and Keegan Paul, 1978).

[9] Putnam, Hilary. "Models and Reality." *Journal of Symbolic Logic* 5 (1980): 464–82.

[10] Smart, J. J. C. *Philosophy and Scientific Realism* (London: Routledge and Kegan Paul, 1963).

4

Confirmation and the Nomological

FRANK JACKSON AND ROBERT PARGETTER

We all suppose that it is sometimes reasonable to project common properties, to argue from the premise that each member of a sample possesses a certain property to the conclusion that members of a population from which the sample is drawn also have that property. Even the most ardent anti-inductivist in the philosophy classroom can be found arguing in just this way when he buys a bottle of wine, takes an aspirin, or is deciding whether to wear a raincoat.

This fact constitutes a major challenge, for (as everyone knows) instantial confirmation, simple induction, the straight rule—call it what you will—faces serious philosophical difficulties. This paper is concerned with two of these difficulties, namely, the one which arises from the fact that any finite sample exemplifies *indefinitely* many common properties and the one which arises from terms like "grue" and "emerose." We will argue that a reassessment of the role of nomological matters in connection with instantial confirmation leads to a resolution of these difficulties. An important feature of our argument is that our claim about the role of nomological matters is plausible in itself. It is not an *ad hoc* contention justified solely by its role in resolving these difficulties.

The paper divides up as follows. In §1 we outline the selection problem arising from the multiplicity of common properties and note that it is a general one with the difficulty over "grue" as a special case. In §2 we offer arguments against the usual, classical approach to this problem. In §3 we criticize the connection between confirmation and the nomological alleged by writers like Nelson Goodman, P. F. Strawson, Israel Scheffler, and John Pollock[1] by way of introducing the role that we see for the nomological. We argue that this role allows an intuitively plausible treatment of the

Reprinted with permission of the University of Calgary Press from *Canadian Journal of Philosophy* 10 (1980): 415–28.

selection problem outlined in §1. Throughout we will simplify (harmlessly, as far as our purposes go) the discussion by focusing on the qualitative question of the existence of confirmation rather than the quantitative of the degree of confirmation. The difficulties we are concerned with pertain to whether simple induction even makes first base.

1. The Selection Problem

We are all familiar with the point that any finite number sequence can be continued in indefinitely many different ways. Similarly, any finite sample exemplifies indefinitely many common properties and has indefinitely many predicates and open sentences true of every member.[2] Any finite collection of objects always has some common property—even if only of the being red *or* square *or* happy *or* . . . variety—and if there is one common property then there are indefinitely many. Because if each member is *F*, each member is also *F* v *G*, for *all G.*

The difficulty this presents the inductivist is that whenever he wishes to apply his rule to project from a sample he has indefinitely many properties to choose from, *and it matters which he chooses.*

It is important to see precisely why it matters. The difficulty is not merely that simple induction leads to indefinitely many hypotheses being supported. This in itself is not objectionable. The difficulty arises from a necessary condition for simple induction being of any use in practice. Simple induction (SI) is only of use if it allows us to go *outside* a sample, to *project.* But this involves going from objects that are, say, γ_1, to objects that are, say γ_2, where γ_1 and γ_2 are mutually exclusive. The objects *to* which one is projecting must differ in some known way from those *from* which one is projecting. For example, SI would be useless if it debarred one from arguing from *examined* emeralds being green to *unexamined* emeralds being green, for then one would never be able to use SI to "break out" of a sample. It will be useful to have a name for the property-pair which marks off the sample from the things one is projecting to. We will call it the differentiating pair. Thus, when one argues from examined emeralds to unexamined ones, <being examined, being unexamined> is the differentiating pair; and when one argues from the emission spectrum of Helium on Earth to the emission spectrum of Helium on the Sun, the differentiating pair is <being on Earth, being on the Sun and not on Earth>. A special case is where $\gamma_2 = \sim \gamma_1$, that is, the pair is jointly exhaustive as well as mutually exclusive. In these cases it is sufficient to talk simply of the differentiating *property*, that which marks off the sample from the things one is proposing to project about. When you argue from examined to unexamined emeralds, being examined is the differentiating property. For simplicity a number of the examples discussed below are of this kind.

It is, therefore, misleadingly narrow to characterize SI in terms of the step from certain αs being β to all (most, etc.) αs being β. A more general characterization requires explicit mention of the differentiating pair, for we are concerned essentially with the step from "All γ_1 αs are β" to "All γ_2 αs are β."

Now suppose you wish to infer from the nature of a sample of αs which are γ_1 to the nature of αs which are γ_2. It would be absurd to use SI to argue that, as all the sampled αs are γ_1, it is reasonable to infer that the γ_2 αs are γ_1. *That use* of SI leads to inconsistency. Examined emeralds do not support unexamined ones being examined. Hence, when we use SI, we must *select* from the properties common to members of the sample when we project. We cannot project *all* the common properties. The *selection problem* is to state the rules governing such selections from the indefinitely many common properties.

It might seem (momentarily) that this problem is easy to solve. When projecting to the γ_2 αs, project all properties common to the sample of αs except γ_1 or any property entailing γ_1. But, of course, this rule is disastrously equivocal concerning what ought to be projected to the γ_2 αs. Let β be any property common to the sampled αs and which does not entail γ_1, then the rule says you should project β. But if β does not entail γ_1, then neither in general will $\delta =$ df $(\beta \,\&\, \gamma_1) \vee ({\sim}\beta \,\&\, \gamma_2)$; so the rule says equally that you should project δ to the γ_2 αs. But a γ_2 α is δ if and only if it is ${\sim}\beta$.

Nelson Goodman's new riddle of induction is a special case of the selection problem.[3] If β is replaced by "*x* is green," and γ_1 by "*x* is examined by *T*," and γ_2 by "*x* is not examined by *T*," then the above definition of δ becomes that of "*x* is grue" (except that we have "not green" for "blue"). The problem raised by "grue" is that of justifying the selection of "*x* is green" rather than "*x* is grue" for projection from examined emeralds to unexamined emeralds.

Another special case of the selection problem arises from certain disjunctive predicates. Here is an example, modified from Scheffler (*op. cit.*, p. 226). Suppose every member of a certain club who I have met is in the Social Register, then each member I have met is either in the Social Register or has been to Pisa. But it would be absurd to infer that Smith, a member I have not met, who I know not to be in the Social Register, either is in the Social Register or has been to Pisa. Because that would be allowing SI to lend support to Smith's having visited Pisa on obviously irrelevant data. What has gone wrong is that the wrong predicate or property has been selected for projection to Smith.

2. The Classical Approach to the Selection Problem

This has been to attempt a partition of properties or predicates into those to which SI applies and those to which it does not. Certain predicates are exalted as projectible, others are castigated as nonprojectible. As Goodman puts it, "I consider a theory of confirmation deficient unless it incorporates a general formulation of the distinction between the right (or projectible) predicates and the wrong ones."[4]

The grounds offered for the distinction vary. For example, sometimes the projectible predicates are said to be projectible because they are entrenched or simple,[5] or because they are purely qualitative[6] or purely ostensive;[7] sometimes because they pick out natural kinds;[8] and sometimes because of the meaning of the associated concepts.[9] There are well-known difficulties for each of these accounts. We will argue however that the classical approach, whatever the fine detail of its manifestation, faces two general difficulties.

The first is that one and the same predicate can be projectible in one set of circumstances but non-projectible in another. One should look, that is, to peculiarities in the circumstances, not to peculiarities in the predicates for a resolution of the selection problem. These circumstances, we will see, are "local" ones, and do not bear, for instance, on the relative entrenchment of the predicates. The second difficulty is that projectibility is relative in a way not adequately acknowledged by the classical approach.

To start with the first difficulty, it is easy to describe circumstances in which even such a classically out-of-favor predicate as "grue" is obviously the correct (and "green" the incorrect) predicate to select to project to an unexamined emerald.

It is a contingent fact about our world that the color of emeralds is not affected by examining them. Emeralds might have been, like chameleons, much more variable in color than they in fact are. In particular, the molecular structure of emeralds might have been such that, though they are naturally blue, the light involved in examining them turns them green. If this were known true of all examined emeralds, then they would be both green and grue, as they are in fact the way things actually are; but the right predicate to project to an unexamined emerald would obviously be "grue," not "green." Because the right thing to believe about an unexamined emerald would be that it was like the examined ones in being naturally blue, and so was blue and not examined, that is, grue.

As the point is important, it is worth laboring it by means of an additional case, namely, the case of the villagers and Caligua. Suppose Caligua enters a village and notices that every villager he meets is pale in the face, and consider the predicate "*x* is palemet" = df "*x* is pale in the face and met by Caligua or *x* is not pale in the face and not met by Caligua." What should Caligua believe about the villagers he has not met—in particular, should he project being palefaced or being palemet to the unmet villagers? It seems obvious that Caligua should not attempt to answer this question by reflecting on which predicate is simpler or ostensibly definable or entrenched or whatever. He should, rather, attempt to answer it by considering the (local) circumstances; more especially, by considering whether these are such that the villagers he has met are only pale in the face because of meeting him, or whether they are naturally pale in the face. If the former, he should project being palemet and so believe that unmet villagers are not pale in the face; if the latter, he should project being pale in the face to the unmet villagers.[10]

The second difficulty for the classical solution is that it fails to reflect the fact that whether or not a predicate ought to be projected is *relative* to the differentiating pair in question, that is, to what knowingly marks off the sample from the object or objects to which projection is being contemplated. Whether or not it is correct to select "grue" for projection depends, in addition to the circumstances discussed immediately above, on whether one is projecting from examined to unexamined emeralds. We all agree that in the normal case (where the color is independent of being examined) it is wrong to project "grue" from an examined to an unexamined emerald. But there is nothing wrong with projecting "grue" to an *examined* emerald. It is perfectly reasonable to expect, on the basis of a number of examined emeralds being grue (and so, green), that a further examined emerald will be grue (and so, green). What is wrong is to expect an *unexamined* emerald to be grue.

Likewise with the modified Scheffler case. There is nothing wrong with projecting being either in the Social Register or having been to Pisa to the next member of the Club who you meet. Because it is reasonable to expect (or raise one's expectation) that the next member of the Club who you meet will be in the Social Register, and so by elementary logic that he will be either in the Register or have been to Pisa. What is wrong is to expect it of the next member of the Club you meet who you know is *not in the Social Register.* But a predicate is no less simple, no less ostensibly definable, no less constitutive of a natural kind and so on, in virtue of being applied to an unexamined emerald or to a person who is not in the Social Register. In short, this aspect of the selection problem is not reflected at all in the classical, castigation-of-predicates approach to it.

We conclude, then, that any adequate treatment of the selection problem must explain both the role of circumstances and the relation to the differentiating pair. We will argue in the next section that our approach via the nomological to the selection problem does this.

Donald Davidson has argued[11] that Goodman tried for too simple a solution to the problem raised by "grue" et al. in that he sought a relevant property (entrenchment) of predicates taken one by one instead of a relation between them. If we are right, the matter is more complex still, doubly so. We need to look at the interrelations between not just α and β, but between them and γ_1 and γ_2; and further at how this complex relation is related to the circumstances. Indeed we take this conclusion to be *comparatively* noncontroversial. You can accept it while denying vehemently our nomological spelling out of it to follow.

In addition to the two general difficulties, the classical response of demarcating the projectible predicates from the non-projectible ones fails to satisfactorily handle relatively simple cases like that modified from Scheffler involving disjunctive predicates. We cannot explain what goes wrong in such cases by ruling that such disjunctive predicates are non-projectible, for that would lead to the position that universals with such predicates in their consequents cannot be instantially confirmed, which is extremely implausible. It is reasonable for scientists to take instances of pulsars which are either white dwarfs or neutron stars to confirm "All pulsars are white dwarfs or neutron stars"; hence, the hypothesis and the disjunctive predicate "*x* is a white dwarf or a neutron star" must be reckoned projectible by proponents of the classical response.

John Pollock suggests a way out of this difficulty for the classical response.[12] He suggests we embrace the (he grants, initially implausible) consequence that instantial confirmation of hypotheses of these kinds with disjunctive consequents is impossible, precisely because they are non-projectible, and then attempts to ameliorate matters by pointing to the possibility of *indirect* confirmation. He suggests that we may indirectly confirm such a (in his view, nonprojectible) hypothesis by instantially confirming projectible hypotheses from which it can be deduced. For example, we might divide pulsars into two mutually exclusive and jointly exhaustive classes and then separately instantially confirm the hypotheses that all members of the first are white dwarfs and that all members of the second are neutron stars. From these two directly instantially confirmed hypotheses we could then deduce, and so indirectly confirm, that all pulsars are either white dwarfs or neutron stars.

This kind of indirect confirmation is not, however, always available. Frequently, both in scientific and everyday contexts, we are able to determine that something is either *F* or *G,* but unable to determine which it is—"He is either a fool or a knave, but I do not know which"—and Pollock's treatment leaves us in the very implausible position that we are unable to instantially confirm hypotheses with disjunctive consequents when this is the case. For example, astronomy might not have developed sufficiently for us to determine whether a celestial object was a white dwarf or a neutron star; all we could tell was that it was one or the other. Surely this would not have debarred astronomers from instantially confirming "All pulsars are white dwarfs or neutron stars" by observing many instances of pulsars which are either white dwarfs or neutron stars, despite their being unable to instantially confirm stronger universals from which this could be deduced.

3. The Role of the Nomological

The idea that appeal to nomological connections may help in handling some of the difficulties of SI is far from new. The usual idea is that instantial confirmation is necessarily confirmation of laws or, more precisely, of the law-like. As Scheffler puts it, "laws may be described as (true) generalizations capable of receiving confirmatory support from their positive instances, in this respect unlike (true) 'accidental' generalizations . . . which do not gain in credibility with each known positive instance."[13]

It would be nice to be able to discuss Scheffler's view in terms of the standard theory of nomological universals and the law-like. Likewise later, when we make mention of subjunctive and counterfactual conditionals, it would be nice to use the standard theory for them. But there is no standard theory in either case. We take it that we have reasonably clear intuitions about both (and will in any case use accepted examples), and that they are connected, perhaps loosely, in the familiar way. We will proceed on this basis, as perforce must everyone.

Scheffler's doctrine is implausibly strong, as others have noted.[14] In any strict sense of "law" very few of the generalizations we confirm in everyday life are laws. Surely we have (nonexhaustive) instantial confirmation of "All wines labeled 'Sauterne' are sweet whites," without having to suppose that this is a law of nature. One of Goodman's own examples is the hypothesis "All men in this room are third sons."[15] He contends that this evidently non-law-like hypothesis is not confirmed by positive instances (short of an exhaustive set including all the men in the room). This is very hard to believe. Suppose the room contains one hundred men and suppose you ask fifty of them whether they are third sons and they reply that they are; surely it would be reasonable to at least increase somewhat your expectation that the next one you ask will also be a third son. Again, surely "All the coins in my left pocket are francs" supports "All the coins in my right pocket are francs." But the relevant universals are notoriously not law-like.

Sometimes the doctrine that confirmation is necessarily confirmation of the law-like appears to be thought an immediate consequence of the idea that (non-exhaustive) confirmation involves *projection*. For example, Pollock observes that such confirmation of "All As are B" must be taken to confirm "If we *were* to encounter another A, it *would* be B." Otherwise we would not be passing *from* the sample to the population in any significant sense.[16] He then infers from the subjunctive nature of this conditional that we are here dealing with the law-like. But there is a crucial difference between the two subjunctive conditionals: "If *x* were an A, *x* would be B" and "If we were to encounter an *x* that is an A, we would encounter an *x* that is B." If we are entitled to project from a sample, we will indeed have support for the latter kind of conditional (except in special cases not relevant here), but the latter kind of conditional, by contrast with the former, is not a mark of the law-like. "All men in this room are third sons" is not law-like but does sustain "If we were to encounter a man in this room, we would encounter a third son."

Nevertheless, there seems to us to be an important insight lying behind attempts to link instantial confirmation and the nomological. As we note in effect in §1, confirmation is of immediate interest inasmuch as it allows one to go beyond one's initial data. Certain αs being β confirming all αs being β is of interest to the extent that it confirms the universal by confirming certain *other* αs being β. But if the sample of αs are all β *purely by accident,* how could this rationally lead one to suppose a continuation of the connection between α and β outside the sample? We urged that a sample of third sons might (weakly) confirm everyone in the room being a third son, despite the latter's evidently non-law-like nature. But in taking such a *sample* to so confirm, you take it that its existence is not a fluke, that there is some reason for the third sons in the sample being in the room. You cannot take its existence both to confirm and to be a fluke.

The important insight then is that one can have no reason to expect the purely accidental to continue. This insight, however, pertains to the sample, not the universal (which after all is what is at issue, and so something one cannot, without begging the question, make assumptions about). What is required is that we bolster SI by imposing a nomological condition on the sample, namely, that the connection between α and β in the sample be nomological with respect to the differentiating pair $<\gamma_1, \gamma_2>$, in the following sense: the members of the sampled things would still have been both α and β even if they had been γ_2 instead of γ_1.

We saw in §1 that SI should be characterized as the inference from "All γ_1 αs are β" to "All γ_2 αs are β." The condition we impose on this inference is that it be reasonable to believe that if the things which are α, β, and γ_1 had been γ_2 instead of γ_1, they would still have been α and β.

Two points should be immediately noted about this condition. First, it is much weaker than Goodman's. "All the coins in my left pocket are francs" supports "All the coins in my right pocket are francs." And our nomological condition is satisfied. We know that the francs in my left pocket would still have been francs and not, say, dimes, had I placed them in my right pocket instead. But the universals are not law-like. Second, the requirement that the γ_1 αβ s would still have been αβs even if γ_2, is a requirement on the γ_1 αβs. It pertains to what we know about the nomic connections between *their* properties, *not* to what we know about the γ_2 αβs. We are not begging the question about the nature of the latter.

This condition (henceforth, the nomological condition) is intuitively plausible. When we use a sample of αs all being β to argue for another α (say) being β, we are taking the β nature of the sampled αs to make it reasonable to believe that an unsampled α is also β. We are, that is, supposing a link between them. But they are known to be different at least in respect of the differentiating pair <γ_1, γ_2>—the sampled αs are γ_1, the other α is γ_2. Hence, when arguing from one to the other in respect of being β, you must suppose that this difference does not matter. But clearly this difference does matter if the sample would not have been of αs all being β if it had been γ_2 rather than γ_1. In brief, in supposing that certain αs being β indicates something about other αs, you must at least suppose that what differentiates the certain αs from the others is not responsible for them being αs which are all β.

We argued in §1 that a satisfactory solution to the selection problem must (i) focus on the circumstances and (ii) reflect the relativity to the differentiating pair. A solution based on the nomological condition does both. According to such a solution, you select for projection from a sample differentiated by being γ_1 rather than γ_2, properties which the sample would have had even had it been γ_2; and obviously which properties these are will depend both on the circumstances and on γ_1 and γ_2.

Consider, for instance, a sample of green emeralds differentiated by being examined. They are all both green and grue; which do we project? According to the nomological condition, we project to unexamined emeralds what would have been the case with the sampled emeralds had they not been examined. Hence, in the circumstances that normally obtain, we should project being green, because we know that in normal circumstances the sampled emeralds would still have been green even if they had not been examined. But in the abnormal circumstances mentioned in §1, where examining the sampled emeralds was known to turn them from blue to green, if the sampled emeralds had not been examined they would have been grue and not green. The nomological condition thus implies, correctly, that in such abnormal circumstances we should project being grue.

The nomological condition also gives the intuitively correct answers when applied to the case, modified from Scheffler, where every member of a certain club who I have met is in the Social Register. We observed that in such a case we would not normally project being either in the Social Register or having been to Pisa to an unmet member of the club who is known not to be in the Social Register. In this case the differentiating property includes being in the Social Register along with being unmet, hence it is only legitimate to project being either in the Social Register or having gone to Pisa to the member who is not in the Social Register if we know that even if the sampled members had not been in the Social Register, they would have either been in the Social Register or have gone to Pisa. But this would, in turn, only be the case if they would have been to Pisa if they had not been in the Social Register. Normally this will not be the case. Properties like being in the Social Register and visiting Pisa are normally quite unrelated.

But, of course, in special circumstances they might be appropriately related. Perhaps I know that every member who I have met was a beneficiary in a will drawn up by an eccentric millionaire, and that a condition of receiving their legacy was that if they were not in the Social Register, they must visit Pisa. In these circumstances I can be confident that if they had not been in the Social Register, they would have gone to Pisa; and so that they would have been in the Social Register or have gone to Pisa if they had not been in the Social Register. Hence, we would then, according to the nomological condition, be entitled to project being in the Social Register or having been to Pisa to a club member who was known not to be in the Social Register. And, in the special circumstances described, this is surely right.

Finally, the nomological condition tells Caligua what he must decide before he projects being pale in the face to the villagers he has not met. He must decide whether the villagers he has met would have been pale in the face if they had not met him. If they would, he may project being pale in the face; but if not, obviously he may not. In the latter case he should rather project being palemet, and so believe of the other villagers that they have normal complexions.

The nomological condition also handles the difficulty for SI posed by terms like "emerose." An object is an emerose if it is either an emerald and examined or a rose and not examined; a sample of examined green emeralds is thus equally a sample of examined green emeroses. But we do not normally expect unexamined emeroses, that is, unexamined roses, to be green on this account. The nomological condition tells us why not. The differentiating property is being examined, and we know that the sample of examined emeroses would not have been emeroses if they had not been examined, for then they would have been unexamined emeralds and unexamined emeralds are not emeroses.

There are, that is, two ways an application of SI can fail to meet the nomological condition. A γ_1 sample of $\alpha\beta$s can fail to bear on γ_2 αs being β because the sampled γ_1 objects would not have been α if they had been γ_2, or because they would not have been β if γ_2. Being grue illustrates the latter, while being an emerose illustrates the former. There are, therefore, really two selection problems: one, emphasized in §1, of suitably selecting β, and the other, brought out by terms like "emerose" of suitably selecting α. In other words, when presented with a sample of objects, there are two connected questions related to SI that one has to face. What kinds of things (emeralds, club members, or whatever) can I project to? and, What properties (being green, being in the Social Register, or whatever) can I project to these kinds of things? And it is all to the good that our nomological condition treats both together.

The selection problem is not so much that a sample supports, by SI, indefinitely many hypotheses; rather it is (as we saw) that SI combined with the role of the differentiating pair—without which SI is useless in practice—leads both to absurd results and to inconsistency. It is absurd (given our knowledge of the independence of color and examining) that examined emeralds being green should support unexamined ones being not green. It is inconsistent that we should have support for unexamined emeralds being both green and not green. We have just been arguing that the nomological condition gives plausible results in practice, that it avoids the absurd results. It can also be shown generally that it avoids inconsistency, given a plausible assumption about counterfactuals.

Let a γ_1 sample of $\alpha\beta$s be also a sample of $\alpha\beta^*$s, where it is impossible for a γ_2 α to be both β and β^*, and suppose projection of one of β and β^* to a γ_2 α is being contemplated. The nomological condition requires that we judge whether we know either "If the sampled objects had been γ_2 instead of γ_1, they would have been γ_2 $\alpha\beta$s" or "If the sampled objects had been γ_2 instead of γ_1, they would have been γ_2 $\alpha\beta^*$s" to be true. And it is plausible that they *cannot* both be true, for they are counterfactuals with identical consistent antecedents and inconsistent consequents.

It has been argued[17] that SI cannot stand in complete theoretical isolation. The mere, unadorned fact that certain αs are βs cannot *in itself* sustain any inference about other αs being β. Our insistence on the nomological condition is in full accord with such a position, indeed the condition specifies one aspect of the required theoretical involvement, for to hold that α and β are nomologically independent of $\langle\gamma_1, \gamma_2\rangle$ is to accept certain theories about the things in the sample and about the interconnections between α, β, γ_1, and γ_2.

The details here are, of course, highly controversial; and we do not wish to enter into them now beyond remarking that if we are right about SI requiring the nomological condition (at least), it is unlikely that the classical problem of justifying induction is essentially the same as that of justifying

SI, contrary to the usual assumption made by advocates of the so-called linguistic justification of induction.[18] In order to justify SI one will need an account of how we know the nomological facts required by the nomological condition, and it is implausible that "meta" applications of SI, each with its own nomological condition to be satisfied, will be adequate to the task.

Our general thought can be simply summarized as follows. Induction is essentially concerned with the projection of similarities across differences. Groups of objects known similar in some ways are hypothesized to be similar in further ways—objects known to be alike in being emeralds are hypothesized to be also alike in being green. But the groups are also different in some ways—some are examined and some are not, for example. There are indefinitely many similarities and indefinitely many differences.

The selection problem is *which* similarities to project across *which* differences. For instance, assuming normal circumstances, grueness may not be projected from the examined before T to the unexamined before T, though it may be projected across other differences. Our answer is to project the similarities you are entitled to believe nomologically independent of the differences (in the precise sense formulated above). Which these are depends on known circumstances and the similarities and differences in question.

Notes

This paper extends and modifies ideas first floated in §III of Frank Jackson, "Grue," *Journal of Philosophy* 72 (1975), 113–31. We are indebted to many discussions with friends and colleagues, including Terry Boehm, Len O'Neill, David Stove, and, particularly, Brian Ellis.

1. In, e.g., Nelson Goodman, *Fact, Fiction, and Forecast,* 2nd ed., New York, 1965; Israel Scheffler, *The Anatomy of Inquiry,* New York, 1963; John Pollock, *Knowledge and Justification,* New Jersey, 1974; P. F. Strawson, *Introduction to Logical Theory,* London, 1952.

2. It does not matter for the arguments of this paper whether it is, strictly, properties, predicates, or open sentences that are projected when you employ simple induction.

3. For arguments against other interpretations of Goodman's new riddle see "Grue," op. cit., §I and §II.

4. Goodman, "Comments," *Journal of Philosophy* 63 (1966), 328–31, p. 329. The classical approach is complicated by the need to detail the connection between projectible predicates and projectible conditionals and hypotheses, see, e.g., D. Davidson, "Emeroses by Other Names," *Journal of Philosophy* 63 (1966), 778–80, and Pollock's comments in *Knowledge and Justification,* op. cit., p. 236. But our discussion is independent of these details, and for instance could, with a slight increase in complexity, have been directed at

classical responses which seek to handle the selection problem by dividing hypotheses into the projectible and the nonprojectible.

5. See, e.g., Goodman, *Fact, Fiction, and Forecast,* op. cit., ch. 4. Goodman sees entrenchment as the basic notion which underlies our judgments of relative simplicity, see his "Safety, Strength, Simplicity," *Philosophy of Science* 28 (1961).

6. See, e.g., Rudolf Carnap, "On the Application of Inductive Logic," *Philosophy and Phenomenological Research* 8 (1947), 133–47.

7. See, e.g., W. C. Salmon, "On Vindicating Induction," *Philosophy of Science* 30 (1963), 252–61.

8. See, e.g., W. V. Quine, "Natural Kinds," in *Ontological Relativity,* New York, 1969, pp. 114–38, and G. Schlesinger, *Confirmation and Confirmability,* Oxford, 1974, ch. 1.

9. See Pollock, op. cit., ch. 8.

10. For other examples illustrating the same general point see "Grue," op. cit., and M. Kelley, "Predicates and Projectibility," *Canadian Journal of Philosophy* 1 (1971), 189–206.

11. D. Davidson, op. cit., see also Pollock, op. cit.

12. Op cit., p. 237. Similar points apply to troubles arising from hypotheses with disjunctive antecedents, but we will not detail them here.

13. Op. cit., p. 225. See also the papers cited in n. 1. Incidentally, Goodman sometimes appears to make the doctrine true by definition, see, e.g., *Fact, Fiction and Forecast,* op. cit. But this, of course, leaves the substantive issue untouched merely effecting a redescribing of it in terms of the adequacy of such a definition.

14. See e.g., A. Pap, "Disposition Concepts and Extensional Logic," in *Minnesota Studies in Philosophy of Science,* Vol. II, ed. H. Feigl et al., Minnesota, 1958.

15. *Fact, Fiction, and Forecast,* op. cit., p. 73.

16. "Laying the Raven to Rest," *Journal of Philosophy* 70 (1973), 747–54, see p. 750, our italics.

17. See, e.g., B. D. Ellis, "A Vindication of Scientific Inductive Practices," *American Philosophical Quarterly* 2 (1965), 1–9.

18. "So-called" because such attempts at justification are best seen not as claims about word usage but as alleging a logical interdependence between SI and the concept of rationality. See, e.g., Pollock, *Knowledge and Justification,* op. cit., p. 204.

5

Induction, Explanation, and Natural Necessity

JOHN FOSTER

I want to examine a possible solution to the problem of induction—one which, as far as I know, has not been discussed elsewhere. The solution makes crucial use of the notion of objective natural necessity. For the purposes of this discussion, I shall assume that this notion is coherent. I am aware that this assumption is controversial, but I do not have space to examine the issue here.

I

Ayer is one philosopher who denies that the notion is coherent. But he also claims that even if it were, it would not help in meeting the problem of induction. "If on the basis of the fact that all the A's hitherto observed have been B's we are seeking for an assurance that the next A we come upon will be a B, the knowledge, if we could have it, that all A's are B's would be quite sufficient; to strengthen the premise by saying that they not only are but must be B's adds nothing to the validity of the inference. The only way in which this move could be helpful would be if it were somehow easier to discover that all A's must be B's than that they merely were so."[1] And this, Ayer thinks, is clearly impossible. "It must be easier to discover, or at least find some good reason for believing, that such and such an association of properties always does obtain, than that it must obtain; for it requires less for the evidence to establish."[2]

Despite its initial plausibility, Ayer's reasoning is fallacious. The first point to notice is that there is a form of empirical inference which is rational, but not inductive in the relevant sense. In the relevant sense, we make an inductive inference when, from our knowledge that all the examined *A*s are *B*s, we infer that all *A*s are *B*s or that some particular unexamined *A* is *B*.

From *Proceedings of the Aristotelian Society* 83 (1982–83): 87–101. Reprinted by courtesy of

In such cases the inductive inference is just an extrapolation from the evidence—an extension to all or some of the unexamined cases of what we have found to hold for the examined cases. Not all rational empirical inferences are of this kind. Thus consider the way in which chemists have established that water is H_20. No doubt there is a step of extrapolative induction, from the chemical composition of the samples examined to the composition of water in general.[3] But this is not the only step of inference. For the composition of the samples is not directly observed: it is detected by inference from how the samples respond to certain tests. The rationale for such inference is the explanatory power of the conclusion it yields. The conclusion is accepted because it best explains the experimental findings—at least it does so in the framework of a more comprehensive chemical theory which is itself accepted largely on explanatory grounds. Thus the conclusion is reached not by extrapolation, but by an inference to the best explanation.

Now look again at Ayer's argument. Ayer is assuming that since "All *A*s must be *B*s" makes a stronger claim than "All *A*s are *B*s," it is no good, in the face of the skeptic's challenge, trying to justify an acceptance of the second via an inference to the first. Now if extrapolative induction is the only form of inference available, Ayer is clearly right. An extrapolation to the stronger conclusion (which associates *A* and *B* across all nomologically possible worlds) already includes an extrapolation to the weaker (which associates *A* and *B* in the actual world) and hence cannot serve to mediate it: any skeptical objection to the smaller extrapolation is automatically an objection to the larger. But suppose we could reach the stronger conclusion by an inference to the best explanation. This would allow an inference to the stronger conclusion to be what justifies an acceptance of the weaker. For it might be precisely because the stronger conclusion is stronger that it has the explanatory power required to make it worthy of acceptance, and thus precisely because we are justified in accepting the stronger conclusion on explanatory grounds that we are justified in accepting the weaker conclusion it entails. This is the possibility which Ayer has missed and which I want to examine in the subsequent sections.

II

Let us focus on a particular case. Hitherto (or so I shall assume), as far as our observations reveal, bodies have always behaved gravitationally—and here I use "gravitational behavior" to cover all the various kinds of behavior, such as stones falling and planets following elliptical orbits, which are normally taken as manifestations of gravitational force. On this basis we are confident that bodies will continue to behave gravitationally in future. But is such confidence well-founded? Does the past regularity afford rational grounds for expecting its future continuation? Here is what strikes me as a natural response. The past consistency of gravitational behavior calls for some explanation. For given the infinite variety of ways in which bodies might have behaved non-gravitationally and, more importantly, the innumerable occasions on which some form of non-gravitational behavior might have occurred and been detected, the consistency would be an astonishing coincidence if it were merely accidental—so astonishing as to make the accident-hypothesis quite literally incredible. But if the past consistency calls for some explanation, what is that explanation to be? Surely it must be that gravitational behavior is the product of natural necessity: bodies have hitherto always behaved gravitationally because it is a law of nature that bodies behave in that way. But if we are justified in postulating a law of gravity to explain the past consistency, then we are justified, to at least the same degree, in expecting gravitational behavior in future. For the claim that bodies have to behave gravitationally entails the weaker claim that they always do. Consequently, our confidence that bodies will continue to behave in this way is well founded. The past regularity does indeed, by means of an explanatory inference, afford rational grounds for expecting its future continuation.

This is just one case. But it illustrates what is arguably a quite general solution to the problem of induction—a solution which is summarized by the following three claims:

(1) The only primitive rational form of empirical inference is inference to the best explanation.

(2) When rational, an extrapolative inference can be justified by being recast as the product of two further steps of inference, neither of which is, as such, extrapolative. The first step is an inference to the best explanation—an explanation of the past regularity whose extrapolation is at issue. The second is a deduction from this explanation that the regularity will continue or that it will do so subject to the continued obtaining of certain conditions.

(3) A crucial part of the inferred explanation, and sometimes the whole of it, is the postulation of certain laws of nature—laws which are not mere generalizations of fact, but forms of (objective) natural necessity.

How this solution works out in detail will, of course, vary from case to case. Sometimes the nomological postulates will form the whole explanation (when the past regularity is a consequence of the laws alone) and sometimes only part (when the past regularity is only a consequence of the laws together with the obtaining of certain specific conditions). Sometimes the predictive conclusion deduced from the explanation will be categorical ("the regularity will continue") and sometimes hypothetical ("the regularity will continue if such and such conditions continue to obtain"). And most importantly, in any particular case, our choice of the *best* explanation will depend to a large extent on what other explanatory theories we have already established or have good reason to accept.

To gain a better understanding of the proposed solution—let us call it the *nomological-explanatory solution* (NES)—three points must now be noted.

(i) Some philosophers hold, contrary to claim (1), that extrapolative induction is a primitive form of rational inference (i.e., *inherently* rational) and that consequently any attempt to justify it by its reduction to other forms of inference is misconceived. I cannot accept this view. Suppose (perhaps *per impossibile*) we knew that there were no laws or other kinds of objective constraint governing the motions of bodies and thus had to interpret the past consistency of gravitational behavior as purely accidental. Such knowledge is logically compatible with the belief that the regularity will continue. But clearly we would have no grounds for thinking that it will. For in knowing that there were no constraints, we would know that on any future occasion any form of behavior was as objectively likely as any other, and this would deprive the past consistency of any predictive value. This result is in line with NES, since the envisaged knowledge explicitly blocks the explanatory inference: if we know there are no laws, we cannot offer a nomological explanation of the past regularity. But how can the result be explained by those who hold induction to be inherently rational? Why should the envisaged knowledge undermine the extrapolative inference unless the rationality of that inference depends on some further inference with whose conclusion the knowledge logically conflicts? I can see no answer to this.

(ii) As indicated in (3), the postulated laws are forms of objective natural necessity. This is crucial. If the laws were mere factual generalizations, or such generalizations set in the perspective of some attitude we have toward them,[4] they would not be explanatory in the relevant sense. In particular, their postulation could not be justified by an inference of a non-extrapolative kind. Thus suppose we construed the law of gravity as merely the fact that bodies always behave gravitationally. There is, I suppose, a sense in which the postulation of this "law" might be taken to explain the past consistency of gravitational behavior—the sense in which to explain a fact is to subsume it under something more general. But it cannot be this sort of explanation which is involved in NES. For if it were, the inference to it would be an ordinary step of extrapolative induction and hence vulnerable to the skeptic's attack. In subsuming the past regularity under a universal regularity we would not be diminishing its coincidental character, but merely extending the scope of the coincidence to cover a larger domain. And it is just this kind of extension which the skeptic calls in question. The reason we can hope to do better with laws of a genuinely necessitational kind is that, arguably, their postulation can be justified by reasoning of a quite different sort. Thus, arguably, we are justified in postulating a law of gravity, as a form of objective natural necessity, because it eliminates what would otherwise be an astonishing coincidence: it enables us to avoid the incredible hypothesis that the past consistency of gravitational behavior, over such a vast range of bodies, occasions, and circumstances, is merely accidental.

(iii) It may be wondered whether past regularities really do call for explanation. Suppose I toss a coin 1000 times, randomizing the method and circumstances of the tossing from occasion to occasion, and each time it comes down heads. Let *H* be the hypothesis that the coin is unbiased, i.e., (in effect) that, for an arbitrary toss, its chances of heads and tails are equal. On the supposition of *H,* the antecedent probability of the run of heads was astronomically small: $1/2^{1000}$. But while astronomically small, it was no less than the antecedent probability of any other of the possible sequences of outcomes: for 1000 tosses, there are 2^{1000} possible sequences and on the hypothesis of no bias each has the same probability. This may lead us to suppose that the occurrence of the run does not count as evidence against *H* and hence does not call for any explanation. For it seems that on the supposition of *H* we should be no more surprised at the run of heads than at any other sequence which might have occurred. In the same way we may be led to suppose that the past consistency of gravitational behavior calls for no explanation—that on the supposition of no laws or constraints this consistency should seem, in retrospect, no more astonishing than any other determinate sequence of behavioral outcomes.

However, this reasoning is fallacious. Suppose I selected the coin at random from a bag of coins, knowing that half are unbiased and half are very strongly biased in favor of heads. Prior to the series of tosses, I could assign equal epistemic probabilities to *H* and to the alternative hypothesis *(H′)* that the coin is heads-biased. If the reasoning above were sound, the subsequent run would not alter these probabilities: that is, even after the run I should have no more reason to accept *H′* than *H,* since on the supposition of *H* the antecedent probability of the run was no smaller than that of any other possible sequence. But this is clearly wrong. Obviously I have very strong grounds for accepting *H′*. If I were to make a habit of betting on *H′* in such circumstances, I could expect to win almost every time. For what matter here are not the relative antecedent probabilities of alternative sequences on the supposition of *H,* but the relative antecedent probabilities of the run on the alternative hypotheses. What makes *H′* overwhelmingly more credible given the evidence of the run is that, antecedently, the run was overwhelmingly more probable on the supposition of *H′* than of *H.* Another relevant factor, of course, are the relative epistemic probabilities of *H* and *H′* prior to the evidence of the run. Had we set the initial probability of *H* higher than *H′*, this would have reduced the strength of the subsequent grounds for accepting *H′* on the evidence. But to make any practical difference, we would have had to set the initial probability of *H′* astronomically low, simply because of the extreme difference in the antecedent probabilities of the run on the two hypotheses.

Let us now apply these considerations to the gravitational case. One hypothesis (H_1) is that, in the respects which concern us, the behavior of bodies is not subject to any laws or constraints, so that any consistent pursuit of gravitational behavior would be purely accidental. What makes the past consistency count so strongly against H_1 is not just that its antecedent probability would be astronomically small on the supposition of H_1 (for this would be true of each possible sequence of behavioral outcomes), but that there are alternative hypotheses on which this probability would be substantially higher and which do not, on the face of it, have a sufficiently lower initial probability to balance this difference. In particular, there is the hypothesis (H_2) that it is a law of nature that bodies behave gravitationally. On this hypothesis, which has been proposed as the best explanation of the consistency, the consistency would be antecedently inevitable. As far as I can see, the only way in which one could rationally retain H_1 in the face of the evidence would be by maintaining that the very notion of natural necessity is incoherent. This is an arguable position (though I think it is mistaken), but, as I said at the outset, I am discounting it for the purposes of the present discussion.

III

NES is beginning to look very plausible. However, there are two major objections to it—in effect, two versions of a single objection. I shall consider one in this section and the other in the next.

The past consistency of gravitational behavior would indeed be an astonishing coincidence if it were merely accidental. Let us agree, then, that we are justified in taking it to be the product of natural necessity: bodies have always behaved gravitationally, within the scope of our observations, because they had to. But why should we suppose that this natural necessity holds constant over *all* bodies, *all* places, and *all* times? Why should we suppose that there is a *universal* law of gravity rather than one which, while covering our data, is restricted in scope to some particular set of bodies or some particular portion of the space-time continuum? For example, with *t* as the present moment, consider the following three nomological hypotheses:

> (A) It is a law *for all times* that (alternatively,[5] it is a law that *at all times*) bodies behave gravitationally.
>
> (B) It is a law *for all times before t* that (alternatively,[5] it is a law that *at all times before t)* bodies behave gravitationally.
>
> (C) (B), and there is no more comprehensive gravitational law.

To justify our belief that bodies will continue to behave gravitationally in future, we have to justify an acceptance of (A) in preference to (C). But how can this be done by an explanatory inference? For both (A) and (C), by including (B), account for the gravitational regularity so far. It seems that to justify an acceptance of (A) we have to fall back on extrapolative induction, arguing that because gravitational behavior has been necessary hitherto, it is likely to be necessary in future. But if so, we have not answered the skeptical problem. Nor, indeed, do we seem to have made any progress at all. For if we have to resort to induction at this point, we might just as well apply it directly to the past regularity without bringing in nomological explanation at all.

Is this objection decisive? Well it is certainly true that (B), and hence both (A) and (C), offer explanations, in the relevant sense, of the past regularity. But this alone is not enough to sustain the objection. What the objector must show is that, as explanations, (C) is not inferior to (A); or put another way, that (B) is not inferior to (A) as a terminus of explanation. And it is on this point, I think, that the defender of NES has a reasonable case. For it seems to me that a law whose scope is restricted to some particular period is more mysterious, inherently more puzzling, than one which is temporally universal. Thus if someone were to propose (C), our response would be to ask why the fundamental law should be time-discriminatory in that way. Why should *t* have this unique significance in the structure of the universe that bodies are gravitationally constrained in the period up to *t* but not thereafter? Barring the postulation of a malicious demon, these questions are unanswerable: any answer we could receive would only serve to show that the fundamental laws were not as suggested—that there was a deeper explanation in terms of time-impartial laws and a difference, relevant to the operation of these laws, in the conditions which obtain in the two periods. It is because these questions seem pertinent and yet are *ex hypothesi* unanswerable that we are left feeling that, as hypothesized, nature is inherently puzzling and precludes an explanation of our empirical data which is both correct and, from the standpoint of our rational concerns, fully satisfactory. And it is for this reason that, presented with the data (the past gravitational consistency) and the alternatives (A) and (C), we are justified in preferring (A). We are justified in preferring (A) because it is the *better* explanation, and it is the better explanation because, unlike (C), it dispels one mystery without creating another: it dispels the mystery of past regularity without creating the mystery of capricious necessity. For the same reason we are justified in preferring (A) to other hypotheses of a similar kind to (C) such as those which restrict the scope of the gravitational law to some particular set of bodies or some particular region of space.

The objector might reply that I am guilty of double standards. I am claiming that in the case of behavior we should avoid unexplained regularity, while in the case of necessity we should avoid unexplained caprice. What I hold to be problematic is, in the one case, a behavioral uniformity not explained by laws and, in the other, a variation in behavioral constraints not explained by a difference in the relevant conditions. But why should our expectations for behavior and necessity be so strikingly different? If there is no problem in expecting irregular behavior when there are no laws to forbid it, why should there be a problem in building a measure of irregularity into the laws themselves? Conversely, if it is reasonable to expect the laws to be uniform over bodies, space, and time, given no positive evidence against it, why should it not also be reasonable to expect uniformities of behavior without the backing of laws? It seems that I am relying on opposite standards of rationality in the two cases.

Well in a sense I am. But that is just because the cases are quite different. What makes them quite different is that, unlike the concept of behavior, the concept of necessity has some notion of generality built into it. Thus try to imagine a world in which there are no conspicuous uniformities, but in which for each object *x* and time *y* there is a separate law prescribing how *x* is to behave at *y*. In such a world everything that happens has to happen, by natural necessity, but there is no uniform system of necessity, or anything remotely resembling one, which imposes the same constraints on situations of the same kind. Each law is concerned with the behavior of a unique object at a unique time. Now it seems to me that such a world is not possible, not because we cannot conceive of such randomness in behavior, but because we cannot conceive of such singularity in the scope of the laws. And this is not just a trivial point about the meaning of the word "law"—a point which we could avoid by choosing another term. Rather, we cannot make sense of the claim that it is naturally necessary for a particular object to behave in a certain way at a particular time except as a claim which is implicitly more general, concerning how it is naturally necessary for objects of a certain type to behave in situations of a certain kind. This is not to say that we cannot conceive of laws (i.e., natural necessities) which are to some degree restricted by some singular reference. We can, I think, conceive of the law postulated by (C), whose scope is restricted to a certain period. But this is only because the restriction leaves room for enough generality of scope for the notion of law to gain purchase. In itself a singular restriction is something which runs counter to the direction of nomological explanation. This is why we serve the purposes of explanation better, if there is a need for explanation at all, by postulating laws without such restrictions, if we can do so compatibly with our data. And in particular, this is why, given the past consistency of gravitational behavior, we rightly regard (A) as a more satisfactory explanation than (C) or any other explanation of a similarly restricted kind, whether the restriction is to a period, to a region, or to a sample of bodies. None of these considerations which apply to our concept of natural necessity carry over to our concept of behavior. There is no implicit notion of generality in our concept of an object's behaving in a certain way at a certain time. Indeed, our rational expectation is that without the backing of laws the total pattern of behavior will be more or less random, not because there is anything to ensure this, but because there is nothing to ensure regularity and because, if it is left to chance, the probability of any significant regularity is exceedingly small. In short, there is something *a priori* perplexing about an arbitrary restriction in the laws and something *a priori* surprising about a coincidental regularity in behavior.

IV

We must now consider the second major objection to NES. Here again I shall continue to focus on the case of gravity.

Let us agree that the past consistency of gravitational behavior calls for nomological explanation and that, since there is no special reason to impose a singular restriction, this explanation should be in terms of laws which are universal in scope. Even so the skeptic has ample room to maneuver. All he needs is some general description "ϕ," not involving, explicitly or implicitly, any singular reference, such that "ϕ" applies to the circumstances of the past regularity but not, as far as we know, to other circumstances or to those particular circumstances with which we are predictively concerned. He can then claim that the past regularity is adequately explained by the hypothesis:

(D) It is a law that in ϕ-circumstances bodies behave gravitationally.

This postulates a universal law, covering all bodies, places, and times. But it does not entail anything about the behavior of bodies in non-ϕ circumstances, i.e., in precisely those circumstances with which, given the evidence of the past regularity, we are predictively concerned. The skeptic will argue that because (D) adequately explains the past regularity, then, to the extent that (A) goes beyond (D), we have no grounds for accepting (A)—in other words, that we have no grounds, other than inductive, for preferring (A) to the alternative hypothesis, (E), which conjoins (D) with the denial that there is a more comprehensive gravitational law. Obviously, the same objection could be applied to any case in which NES was invoked to justify an inductive inference.

It is not easy to evaluate this objection. One difficulty is that we need some general but reasonably detailed account of what makes one explanation better than another. Clearly there is at least one factor on the side of the skeptic: if two hypotheses both explain the data and one hypothesis is stronger than the other (i.e., entails but is not entailed by it), then, other things being equal, the weaker hypothesis is to be preferred (thus if other things were equal, (D) would be preferable both to (A) and to (E)). What is far from clear is how we are to determine whether other things are equal. It is easy enough to say something very general and non-committal: e.g., other things are not equal just in case the weaker hypothesis, while explaining the data, postulates some state of affairs which itself calls for further explanation of a kind which the stronger hypothesis supplies, or the conjunction of the weaker hypothesis and the negation of the stronger postulates a state of affairs which is inherently more puzzling than the state of affairs which the stronger hypothesis postulates. But what we need, to evaluate the objection, is a set of more specific principles, justified independently of induction, which will enable us to decide case by case whether a state of affairs does call for explanation or is inherently puzzling. And the formulation of such principles would be a large and difficult task, if it is possible at all. A further difficulty, at least for a defender of NES, is that there is an infinite range of non-equivalent descriptions which could play the role of "ϕ." To rebut the objection entirely, it would be necessary to divide this range into a finite number of categories, show that the differences within each category were, relative to the present issue, irrelevant, and then rebut the objection for each category. This, too, promises to be a difficult and perhaps impossible task, even if, for each separate description, the objection could be shown to fail.

All I can do here is to examine some of the more obvious cases on their own merits. One such case would be to model (D) and (E) on the cases of (B) and (C) considered earlier. Thus suppose "S" is a state-description of the universe at *t* (i.e., the present moment) and *"Fx"* is defined as "the universe is in state S at time *x*." Then we have as examples of (D) and (E):

> (D^1) It is a law that at any time before an F-time bodies behave gravitationally.
>
> (E^1) (D^1) and there is no more comprehensive gravitational law.

For all we know, an F-time will not occur in the future. So explaining the regularity by (D^1) provides no basis for extrapolation. The question is: are there non-inductive grounds for claiming that (A) serves better than (D^1) as a terminus of explanation? And this question becomes: is the state of affairs postulated by (E^1) inherently more puzzling than that postulated by (A)? I think the answer is "Yes," for two reasons. Firstly, in effect (E^1) involves action at a temporal distance. For if (E^1) were true, then (and here I assume, for simplicity, that *t* is the only F-time) each past instance of gravitational behavior would directly causally depend, in part, on the intrinsic state of the universe at *t,* there being no continuous causal chain mediating this causal dependence and spanning the temporal interval between *t* and time of the behavior. Secondly, since *t* is subsequent to the past instances of gravitational behavior, the direction of the causal influence involved would be from later to earlier: the occurrence of gravitational behavior in the past would be partly the causal result of the state of the universe now. In both these respects, and especially the second, what (E^1) postulates is inherently more puzzling than what (A) postulates, and consequently (A) is a better explanation than (E^1) and better, as a terminus of explanation, than (D^1).

One way for the skeptic to eliminate both these defects would be as follows. Take each occasion *i* of observed gravitational behavior and form a very detailed description "F^i" of the intrinsic conditions obtaining immediately prior to this behavior—a description sufficiently detailed to distinguish it, as far as we know, from the conditions which will obtain on any future occasion or on those future occasions with which we are predictively concerned. We then let "ϕ" be the disjunction of these descriptions, so that "in ϕ-circumstances" means "either in F^1-circumstances or in F^2-circumstances or . . ." where the disjunctive list exactly covers all the specific conditions in which gravitational behavior has occurred and been detected so far. Let us call (D) and (E) thus interpreted (D^2) and (E^2). Then (E^2) avoids the two mentioned defects of (E^1). It does not postulate any backwards causation or any direct causation at a temporal distance. If (E^2) were true, each past instance of gravitational behavior would causally depend solely on the intrinsic conditions obtaining on that occasion.

Are there any other grounds for claiming that (A) is a better explanation than (E^2) and better, as a terminus of explanation, than (D^2)? I think there are. The first point to notice is that, in effect, (D^2) explains the past regularity of gravitational behavior by providing a separate explanation of each past instance. Because "ϕ" does not signify a natural generic property, but rather a disjunctive list of the complex properties separately drawn from the separate instances, it would be less misleading to reformulate (D^2) as a long list of separate hypotheses: "It is a law that in F^1-circumstances bodies behave gravitationally; it is a law that in F^2-circumstances bodies behave gravitationally." The whole list would provide an explanation of the past consistency only in the sense that each hypothesis provided an *ad hoc* explanation of one behavioral instance. How then should we respond to someone who offers (D^2) as a terminus of explanation, i.e., asserts (E^2)? Well, we are likely to find the state of affairs he postulates inherently puzzling, since the way the laws discriminate between ϕ and non-ϕ circumstances is not based on any natural mode of classification: it seems peculiar that just the listed circumstances should be gravitationally efficacious when they are no more similar to each other than they are to other circumstances. Still, it may be hard to establish that the grounds for this puzzlement are non-inductive, and for this reason I would put the stress on a different point. Even though in a sense (D^2) explains the past consistency of gravitational behavior (by separately explaining each instance), it leaves us with another consistency which calls for explanation and which would be very hard to explain if we accepted (E^2): for although there are infinitely many types of circumstances (all those that are non-ϕ) to which the (D^2)-law does not apply, whenever we have checked for gravitational behavior the circumstances have always been of a type (ϕ) to which the law does apply. This would be an astonishing coincidence if it were purely accidental, and, on the face of it, it would be purely accidental on the supposition of (E^2).[6] No such coincidence arises on the supposition of (A), and for this reason, if no other, (A) is the preferable hypothesis.

These are, of course, only two examples of the way in which the skeptic might pose the objection and the fact that we can rebut them does not mean that he cannot turn to others. But the onus is now on him to produce a convincing case. My guess is that whatever he chooses for "ϕ" there will be some way of vindicating our preference for (A).

V

It goes without saying that this discussion has been sketchy and inconclusive. I have concentrated throughout on a single example of inductive inference (that concerning gravity) and even with respect to this example I have not had space to deal adequately with the issues raised in the last section. Moreover, I have carried throughout the controversial assumption that we can make sense of the notion of objective natural necessity. What I hope I have shown is that, if we can make sense of this notion, the nomological-explanatory solution is worth considering. I, for one, find it quite plausible.

Notes

Meeting of the Aristotelian Society held at 5/7 Tavistock Place, London WC1, on Monday, 31 January, 1983 at 6.00 p.m.

1. Ayer, *The Central Questions of Philosophy* (Pelican 1976), pp. 149–50.

2. Ibid., p. 150.

3. I.e. to what *pre-theoretically* qualifies as water.

4. For an attitudinal account, see Ayer, *The Concept of a Person* (Macmillan 1963), ch. 8.

5. I shall not inquire as to whether these are merely alternative formulations or differ in substance. As far as I can see, if there is a difference in substance, it does not affect my argument.

6. Of course it was not accidental that the skeptic chose to postulate a law which exactly covered the examined cases. But that is beside the point.

6

Armstrong on Laws and Probabilities

BAS C. VAN FRAASSEN

The question of David Armstrong's recent book, *What Is a Law of Nature?* would seem to have little point unless there really are laws of nature. However that may be, so much philosophical thinking has utilized this concept, that an inquiry of this sort was needed whether there are or not. The book begins with a devastating attack on so-called Regularity views of law, and then proceeds with an exposition of Armstrong's own answer to the question. I wish to raise here some difficulties for Armstrong's answer, concentrating on his account of probabilistic laws where I see the severest problems. To locate myself with respect to his approach, however, I shall first enter some lesser demurrals.

1. The Supposed Significance of the Topic

Two motives are found in the literature for the discussion of laws of nature. The first is the existence of arguments to the effect (a) that there must be laws of nature and/or (b) that we must believe that there are such laws. The second lies in a certain form of reflection on science. I have the impression that, although the second motive is displayed more prominently, one cannot share it unless persuaded already by the first.

As to the first motive: both sorts of arguments appear to be based on very general and surely very plausible premises. For (a) the premise is that there are pervasive regularities in nature. It is inferred (and the inference is justified by some appropriate non-deductive rule) that therefore there must be laws of nature. For (b) the premise is that we have knowledge or rational beliefs about the future or the unobserved. It is argued then that this is possible only if we can have knowledge of or rational belief in the reality of laws.

From *Australasian Journal of Philosophy* 65, 3 (1987): 243–60. Reprinted by permission of Oxford University Press.

Peirce's famous essay "The Reality of Thirdness" has still perhaps the most immediate impact, but Armstrong's recent book provides interesting recent examples of such arguments. Given the weakness of the apparent premises, and the exceedingly far-reaching conclusion, these arguments should indeed be central concerns of ontology and epistemology.

The more fashionable motive, however, is the second: the assertion that laws are what the sciences seek to discover. Accordingly, it is concluded, "the nature of a law of nature must be a central ontological concern for the philosophy of science" as Armstrong says in the first section of his book. This motive, thus placed in the foreground, contrasts with the fact that the discussion of laws moves, insofar as it relates to science, on a very abstract level indeed. Armstrong points this out explicitly, and calls it appropriate: "[i]t turns out, as a matter of fact, that the sort of fundamental investigation which we are undertaking can largely proceed with mere schemata of this sort [*It is a law that F*s *are G*s]. . . . To every subject its appropriate level of abstraction" (pp. 6–7). That is fair enough, but we face two possibilities. The first is that the arguments and conclusions at this abstract level are independent of any view on what science is. In that case the motive announced at the outset may be disingenuous. The second possibility is that the reasoning does involve a specific, substantive view of what science is. In that case we must ask whether this view is not itself a dubitable premise.

There is a view of science which the more metaphysical writers on laws appear to share, and which informs their discussion. Armstrong presents it on the first page. Natural science traditionally has three tasks: *first,* to discover the topography and history of the actual universe; *second,* to discover what sorts of things and sorts of properties there are in the universe; and *third,* to state the laws which the things in the universe obey. The three tasks are interconnected in various ways. David Lewis expresses his own view of science in such similar comments as these:

> Physics is relevant because it aspires to give an inventory of natural properties. . . . Thus the business of physics is not just to discover laws and causal explanations. In putting forward as comprehensive theories that recognize only a limited range of natural properties, physics proposes inventories of the natural properties instantiated in our world. . . . Of course, the discovery of natural properties is inseparable from the discovery of laws. (1983; pp. 356–57, 364, 365)

Where does this highly specific view of science come from? If it springs from an independent theory about laws and properties, this view is being put forward against rival views of science to be tested. If it fares well, its success will support the theory of laws and properties from which it derives. That would be very different, however, from a use of the view, taken as independently known to be correct, to motivate acceptance of the concept of law as a central concern for the philosophy of science. For my part, I do not share that view of science, and I have as yet seen no real effort to test it or to argue for its adequacy.

There is a third possible role for the recurring references to science. It may be that some philosophers take it as a regulative ideal for metaphysics, to be so developed that it be possible to view the sciences as among its proper parts or subdisciplines. This would perhaps place such constraints on a metaphysical system as that any entities posited as real should be a proper subject of inquiry for science. Hence there would be a constant need for, at least, accompanying plausibility arguments. The introduction of universals, natural properties, laws, physical necessities would have to be accompanied by the formulation of views on science that place inquiry into what these are like, within the proper reach of science. This would make the study of science necessarily important for the philosopher who posits laws; not, however, the study of laws to the philosopher of science.

It appears, therefore, that to have the second motive for the study of laws, one needs already to have been convinced by the arguments, whose existence provides the first motive.

2. Laws Analyzed in Terms of Universals

The account of laws of nature that Armstrong gives is a more sophisticated version of the ones found in papers by Dretske and Tooley. (See bibliography.) It will be useful by way of introduction to look briefly at their accounts first (with apologies, since they clearly deserve more attention in their own right as well).

It may be tempting, says Dretske, to identify laws with a special class of universal truths, just because alternatives might seem to place the law beyond our epistemic grasp. How could one have evidence, for example, for what is the case in other worlds, or about entities that do not exist in ours? But that tempting identification would fail to do justice to a good half of the intuitions about laws—especially that laws "tell us what (in some sense) must happen, not merely what has [happened] and will happen" (p. 263). The solution he (and Tooley and Armstrong) offer is that a law does say more than the corresponding universal truth, but the "must" is explicable in terms of real parts or aspects of our world. The offered form is very simple: *It is a law that all F*s *are G*s is true exactly if a certain singular statement *Fness* → *Gness* is true. The term *Fness* refers to a universal, what a thing must have in order to be *F*. The arrow stands for a relation between them; Dretske does not introduce a standard name for it, but Tooley calls his variant "nomic necessitation." The connotations of the terms used convey a crucial fact about this relation: the argument from *Fness* → *Gness* to *All F*s *are G*s is valid.[1]

Dretske draws the usual connection between laws and necessity as follows: the argument from *Fness* → *Gness* and *b is F* to *b must be G* is valid as well. By universal generalization one infers: the argument from *Fness* → *Gness* to *All F*s *must be G*s is valid. It is not clear that *It must be the case that all F*s *are G*s can be inferred! How could such an inference be explicated? These writers naturally eschew possible world semantics, even if they are happy to consider possible worlds of various sorts (ways the world could have been) in examples. Of the three, Tooley is most explicitly concerned with semantics; it is not difficult to see in his article (esp. pp. 678–79) the possibility of maneuvers to construe the necessity operator so as to make the above inference valid. I will come back to these questions below.

Tooley is most explicit about another shared motivation. What is or is not a law is independent of almost anything you care to mention—certainly of epistemic and pragmatic factors, but also of form and meaning. Reichenbach contrasted the two universal statements "No signals are faster than light" and "No gold cubes are larger than a cubic mile" (1947, p. 368; 1954, pp. 10–11). We accept the former, but not the latter as a law. The situation would be different, Reichenbach adds, if we had accepted a theory that entails the gravitational collapse of larger gold cubes. Thus universal, purely qualitative statements may or may not be laws. Tooley gives the parallel example (which David Lewis tells me dates back at least to the sixties) of "All spheres of uranium (U235) are less than a mile in diameter," which one could easily enough accept as a law (the critical mass for fission being attained well short of that amount), but not when "uranium" is replaced by "gold." The law/non-law distinction is therefore beyond the reach of syntactic/semantic distinctions.

The addition of facts about the world conjoined to syntactic and semantic features also does not help. Tooley considers two possible worlds. In the first it is a law that all *F*s are *G*s. In the second it is a law instead that an *F* has a chance of 0.9999 of being a *G;* there are actually few *F*s and they are all *G*s. Hence the two worlds do not differ with respect to how many *F*s are *G*s. He also gives the example of a world in which two kinds of particles never come close enough to interact; there is a law which governs their interaction, but equally well, there might not have been. Thus he clearly countenances the idea of worlds which are factually alike but differ in their laws, and does not hold that laws supervene on facts about particulars.

Tooley has a clear overall program for semantics along metaphysical lines. This is "the project of providing an account of the truth conditions of nomological statements in non-nomological terms," part of the larger "more general program of providing truth conditions for intensional statements in purely extensional terms" (p. 672). This sounds innocuous, but it soon appears that Tooley equates finding truth conditions with finding "truth-makers." To the question how there can be fundamental laws without any positive instances, he says "The only possible answer would seem to be that it must be facts about *universals* that serve as the truth-makers" and he elaborates this as follows:

> Facts about universals will consist "of universals" having properties and standing in relations to other universals. How can such facts serve as truth-makers for laws? My basic suggestion here is that the fact that universals stand in certain relationships may *logically necessitate* some corresponding generalization about particulars, and that when this is the case, the generalization in question expresses a law. (p. 672)

The question addressed arises for Dretske as well: how is the statement *Fness* → *Gness* related to *All F*s *are G*s? Dretske says the argument from the former to the latter is valid, and Tooley calls their relationship logical necessitation. But surely it is not a matter of logic at all? The contention was that *It is a law that all F*s *are G*s entails, at least, that all *F*s are *G*s. This claim requires substantiating, and the first step is to equate the law statement with *Fness* → *Gness.* The second step is to claim that this relational statement entails that all *F*s are *G*s. But this claim seems no easier to establish by deductive logic than the former. Armstrong used "necessitates" instead of "→" which is a perspicuous reminder that the argument is meant to be valid. In commenting on this, David Lewis points out correctly the validity is no more guaranteed by the meaning of "necessitates" here than someone is guaranteed to have mighty biceps by the meaning of "Armstrong" (1983, p. 366).

3. Does Armstrong Escape the Problem?

Time now to turn to Armstrong's more developed theory. Armstrong had developed a theory of universals beforehand, and gives an account of laws within that. His world has in it particulars and universals. The particulars come in two sorts, objects and states of affairs. A state of affairs will always involve some universal, whether monadic (property) or dyadic, triadic, . . . (relation). When I say particulars, I mean entities which are not universals, but which instantiate universals. It is countenanced that universals may instantiate other universals, and that there is a hierachy of instantiation. So it is natural, as Armstrong does, to extend the terminology: call an nth order universal one which has instances only of $(n - 1)^{th}$ order, call it also an $(n + 1)^{th}$ order particular. Now call the particulars which are not universals, first-order particulars. When I say "particular" without qualification I shall mean these first-order particulars. If *a, b* are particulars and *R* a relation, and *a* bears *R* to *b,* then there is a state of affairs, *a's* bearing *R* to *b*—Armstrong refers to it as *Rab*—in which these three are "joined" or "involved." If *a* does not bear *R* to *b,* there is no such state of affairs—a state of affairs which does not obtain is not real.

With any theory of universals there is the threat of "third man" problems, so let us look first at his reaction to Bradley's famous regress. In the state of affairs described above, the terms *R, a, b* are all names, one of a universal and two of particulars. These three entities are "joined" in the state of affairs; but how? Is it that there is a certain three term relation *R′* which *R* bears to *a* and *b?* More generally, how are universals related to their instances? If a regress were accepted, it might or might not be vicious. After all, suppose $\mathbb{R}$ is the set of real numbers, and X one of its subsets. Then in set theory we could say $b \varepsilon X$ exactly if $\langle b, X\rangle \varepsilon \{\langle y, Z\rangle : y \varepsilon \mathbb{R} \; \& \; Z \subseteq \mathbb{R} \; \& \; y \varepsilon Z\}$. Calling the latter set, a subset of $\mathbb{R} \times P(\mathbb{R})$, by the name X′ we continue with the equivalent $\langle\langle b, X\rangle, X'\rangle \; \varepsilon X''$ for the membership relation restricted to $(R \times P(\mathbb{R})) \times P(\mathbb{R} \times P(\mathbb{R}))$ and so forth *ad infinitum.* But even if a regress does not lead to a contradiction, that does not mean it is a satisfactory thing to have around. Armstrong adopts a view of the Aristotelian (moderate realism) sort in order to avoid Bradley's regress. If universals were substances (i.e., entities capable of independent existence), he says, the regress would have to be accepted. But they are not substances, though real: they are abstractions from states of affairs. One consequence is that there are no uninstantiated universals. The more important point here is that we should not regard states of affairs as being constructed out of universals and particulars, which need some sort of ontological glue to hold them together.

We cannot embark on the general theory of universals here; we need to consider only those details crucial to Armstrong's account of laws. In this account the symbol "N" is used, *prima facie* in several roles; I shall use subscripts to indicate *prima facie* differences and then state Armstrong's identifications. There is first of all the relation N_1 which states of affairs can bear to each other, as in

(1) N_1(*a*'s being *F*, *a*'s being *G*)

This N_1 is the relation of necessitation between states of affairs; formula (1) is a sentence which is true if and only if *a*'s being F necessitates *a*'s being *G*. But for this to be true, both related entities must be real; therefore (1) entails that these two states of affairs are real, hence

(2) *a* is *F* and *G*

However, both states of affairs could be real, while (1) is false, so N_1 is *not* an abstraction from the states of affairs of sort (2)—for then it would be the conjunctive universal (*F* and *G*).

This relation N_1 has sub-relations, in the sense that the property *being colored* has sub-properties (determinates) *being red, being blue.* One sort is the relation of *necessitation in virtue of the relation(s) between F and G,* referred to as N_1 (*F*,*G*):

(3) N_1 (*F*,*G*) (*a*'s being *F*, *a*'s being *G*)

This is a particular case of (1), so (3) entails (1) and hence also (2), but again the converse entailments do not hold. In (3) we also have a sort of universalizability. Note that (3) says (a) that the one state of affairs necessitates the other, and (b) that this is in virtue solely of the relation between *F* and *G*. Hence what *a* is does not matter. Of course we cannot at once generalize (3) to all objects, for (3) is not conditional; it entails that *a* is both *F* and *G*. Thus we should say that what *a* is does not matter beyond the entailed fact that it is an instance of those universals whose relation is at issue. We conclude therefore that if (3) is true, then for any object *b* whatever, it is also true that

(4) If *b* is (*F* and *G*) then $N_1(F,G)$(*b*'s being *F*, *b*'s being *G*)

Now what is the relation between *F* and *G* by virtue of which *a's* being *F* necessitates *a*'s being *G?* It is the relation of necessitation between universals, Armstrong's version of Dretske's "→" and Tooley's nomic necessitation. Let us call this N_2:

(5) $N_2(F,G)$

Thus if (3) is true then (4) and (5) must be true, and indeed, (3) must be true *because* (5) is true ("the truth-maker"). This N_2, the target of course of the remark by David Lewis which I reported earlier, Armstrong takes as not calling for further illumination:

> the inexplicability of necessitation just has to be accepted. Necessitation, the way that one Form (universal) brings another along with it as Plato puts it in the *Phaedo* (104d–105) is a primitive, or near primitive, which we are forced to postulate. (p. 92)

"Forced to postulate" because the way the actual particulars (including states of affairs) are in our world cannot determine whether (3) is true—and (3) must have, for Armstrong, a truth-maker, and that truth-maker must therefore be the way the non-particulars are.

But is there not another threat of regress here? Consider the claim that all *F* are *G* because *F* necessitates *G*. This is supported by the "deeper" claim that if *F* necessitates *G* then all *F* are *G*. That is if one sort of relation holds between *F* and *G* (necessitation) then another does too (extensional inclusion). Does *this* statement need a truth-maker? Needs there to be a "trans-order law," which entails that the one sort of relation among universals is always accompanied by the other sort?

Armstrong may have a special escape here. According to him, the universal $N_1(F,G)$—a relation between states of affairs—is identical with the state of affairs N_2 (F,G). In that case we can drop the subscripts and say that the state of affairs $N(F,G)$(*Fa*, *Ga*) instantiates the state of affairs $N(F,G)$—in the way that *Rab* instantiates *R*, generally. In that case the question whether $N(F,G)$ entails (If *Fb* then *Gb*), can be approached by logical means. For suppose N(*F,G*) is real, i.e., *F* bears N to *G*. Then it must have at least one instance; let it be described by (3) above. But in that case (4) is also true. We conclude therefore that if N(*F,G*) is real, then for any object *b,* if *b* is both *F* and *G,* then N(*F,G*) (*b*'s being *F, b*'s being *G*).

But we can go no further. What we have determined is this: if there is a law N(*F,G*), then *all* conjunctions of *F* and *G,* in any subject, will be because of this law. There will be no *F*s which are only accidentally *G*. That is an interesting and undoubtedly welcome consequence. Although Armstrong does not give this argument, I must assume that he introduced the curious identification of N_1 and N_2 to reach some such benefit. However, this benefit is not great enough to get him out of the difficulty at issue. For what *cannot be deduced,* from the universal quantification of (4), is that all *F*s are *G*s. Any assertion to that effect must be made independently. Nothing less than a bare postulate will do, for there is no logical connection between relations among universals and relations among their instances.

It seems to me, looking back over this examination of the accounts of laws in terms of universals, that they are not very successful by the authors' own lights. To highlight the most serious problem, imagine a discussion in the Renaissance between a Schoolman (with a complex theory of natures, substantial forms, *composition,* and occult qualities) and a new Mechanist (with a naive theory of atoms of different shapes, with or without hooks and eyes). The Schoolman says that the Mechanist account must eventually rely on the regularities concerning atoms, such as that their shapes remain the same with time, and there must be a *reason* for these regularities in nature. But the Mechanist can reply that whatever algebra of attributes, etc., the Schoolman can offer him, the inference from the equations of that algebra to regularities in the behaviors of atoms, must rest on some further laws which relate attributes to particulars. For example, if A is part of B, it may follow that instances of A are instances of B, but not without an additional premise which justifies, in effect, the suggestive part-whole terminology for the indicated relation between universals. Of course, if we define A to be part of B exactly if it is necessary for instances of A to be instances of B, the argument becomes valid. But then is the necessity appealed to in the definition itself grounded in some further reality? And if not, if there can be a necessity not further grounded in some further reality, he would like to return to his atoms, please, and say that their postulated regularities are not grounded in any further reality—it's just the way atoms are.

4. The Account of Probability in Statistical Laws

With the advent of modern physics, a deterministic world-picture began to dominate science; recently it has given way again to indeterminism. The discussion of probabilistic laws of nature is interesting from several points of view; it may throw light on the structure of statistical theories, vexing enough also for philosophers who do not think about laws of nature. If my conclusions in this section are correct, however, then Armstrong's theory of laws of nature is, as it stands, entirely incapable of explicating the concept of a probabilistic law of nature.

Before turning to Armstrong's account, let us rehearse the simplest sort of case, radio-active decay. The half-life of radium is approximately 1,600 years. This law of atomic physics, which predates quantum theory, means that, given an initial amount of radium, half of it decays in the first 1,600 years, half of the remainder in the next 1,600 years, and so on. (Decay means that it turns into radon.) Unlike Zeno's frustrated traveler, however, the process must come to an end, according to microphysics, because a finite amount of radium consists of a finite number of atoms. This implies at once that the half-life law cannot be true; we can't have half of an odd number of atoms decaying. This applies equally well to the complete law of radioactive decay to which the half-life is a corollary:

(1) $N = Ie^{-At}$

where I is the initial amount, N the amount left after the time interval t, and A is radium's decay constant. But quantum theory says that a radium atom has a probability e^{-At} of remaining stable for interval t (equivalently, a probability $(I - e^{-At})$ of decaying into radon during that interval), with the effect of atomic interaction negligible. Law (1) now holds on the reinterpretation: N is the expectation value of the remainder after interval t. The old law holds with "practical certainty" for "reasonable" choices of I and t. This is a simple example, because it involves only a single parameter, with two values (radium, radon).

Armstrong begins his chapter 9 by asking us to consider an irreducibly probabilistic law to the effect that there is a probability *P* of an *F* being a *G*. I imagine that *G* may include something like: remaining stable for at least a year, or else, decaying into radon within a year. Adapting his earlier notation, he writes

(2) ((Pr:*P*)(*F*,*G*))(*a*'s being *F*, *a*'s being *G*)

Read it, to begin anyway, as *There is a probability P, in virtue of F and G, of an individual F being a G.* As with N, (Pr:*P*)(*F*,*G*) is a universal, a relation, which may hold between states of affairs, but of course only real ones. Suppose now that *a* is *F* but not *G*. Then (2) is not true. So (2), properly generalized, does not say something true about any *F*, but only about those which are both *F* and *G*. That is not what the original law-statement looked like. Nor does Armstrong wish to ameliorate this by having negative universals, or negative states of affairs, or propensities (i.e., properties like *having a chance P of becoming* G which an *F* could have whether or not it became *G*). Thus *a probabilistic law is a universal which is instantiated only in those cases in which the probability is "realized."*

Suppose there is such a law; what consequences does this have for the world? The real statistical distribution should show a "good fit" to the theoretical distribution described in the law. The mean decay time of actual radium should show a good fit to law (1)—on its new, probabilistic interpretation—for example, also on Armstrong's construal. But I don't see why it should. We can divide the observed radium atoms into those which do and do not decay within one year. Those which do decay are such that their being radium atoms in a stable state bears (Pr:e^{-A}) (radium, decay within one year) to their decaying within one year. The other ones have no connection with that universal at all. Now how should one deduce anything about the proportions of these two classes or about the probabilities of different proportions?

Open questions are not satisfactory stopping points, so let us leave the connection with actual frequency aside, and concentrate on probability alone. The reality of (Pr:*P*)(*F*,*G*) has one obvious consequence: a universal cannot be real without being instantiated, so there is at least one *F* which is *G*. Thus we have, for example:

(3) If it is a law that there is a probability of 3/4 of an individual *F* being a *G*, and there is only one *F* then it is a *G*.

This is worrying because a probability of 75% has turned into 100%, but a one-*F* universe is perhaps so unusual that it can be ignored. However, suppose there are two *F*s, call them *a* and *b*. If we ignore the Principle of Instantiation, and assume this is the only relevant law that is real, we calculate:

the probability that both are *G* equals 9/16, the probability that *a* alone (or *b* alone) is *G* equals 3/16, and that neither is *G* equals 1/16. But the Principle rules out the last case, so we must conditionalize on its negation. This means dividing by 15/16, and we deduce, after a few steps:

> (4) Given the law that there is a probability of 3/4 of an individual *F* being a *G*, and *a*, *b* are the two only *F*s, then the probability that *a* is a *G* equals 4/5, and the probability that *a* is a *G* given that *b* is a *G*, is a bit less (namely, 3/4 again).

So the trouble is not confined to a one-*F* universe; it is there as long as there is a finite number of *F*s: if the law says probability *P*, and there are n *F*s, then the probability that a given one will be *G* equals *P* divided by $(1-(1-P)^n)$. For very large n, this is indeed close to *P*, but the difference would show up in sufficiently sensitive experiments. Should we recommend this consequence to physicists, if they have ever to explain apparent systematic deviations from a probabilistic law?

Let us briefly see what this would mean for radio-active decay. I do not know whether it is *remaining stable for interval t*, or rather *decaying into radon within interval t*, that is a universal. If the *former*, we note that e^{-At} is positive for all intervals t, and so there must be a radium atom which remains stable for at least that long, for every t. This means that either there is one which never decays, or else there are infinitely many radium atoms (remaining stable for at least one year, at least two years, at least three . . .). If the *latter*, there must be at each instant either an atom which decays at once or else there must be infinitely many, each decaying within a shorter time interval after that instant than the preceding ones in the list.

The finitude of radium available entails that only the first alternative under the first option is possible. Hence we have deduced the existence of a radium atom which remains stable forever!

We still have the corollary noted at the end of (4). Individual decay is not statistically independent—given that some atoms decay the probability for any given other one decaying is less. This will be true regardless of how the atoms are separated in space and time; so we have here a correlation inexplicable by any causal model. That is certainly typical of atomic physics today, but radio-active decay for spatially separated atoms was not so far a case in point. It is of course independently interesting for those, perhaps more traditional, realists who think that significant statistical correlations must have a causal explanation. It appears that Armstrong's account entails that they are wrong.

Let us return to this account. He proposes that we regard (Pr:*P*) as a subrelation of N, rewriting it as (N:*P*) and identifying N with (N:1). This seems to me to go a bit far because N was used to formulate deterministic laws which brook not even a single violation. Probability *one* statements do. (That is why it was consistent in the paragraph before last to contemplate a single atom which remains stable forever, or which decays at once, both of which have probability zero. Armstrong has indicated interest in a version of probability theory in which, if something ever happens, then it has nonzero probability [which may be infinitesimal]. That would *prima facie* worsen the problem of the paragraph before last.) More problematic is the interpretation of "(N:*P*)." The title of chapter 9, section 2, and most relevant passages suggest that we read this as "the probability of necessitation equals *P*." That means necessitation is the same relationship as before, but the law is not such that if it is real then all *F*s provide instances of it. Because of the difficulties below, we should keep in mind the other possible interpretation, namely that N—like temperature or propensity—has degrees, of which *P* provides a measure.

The main difference I see between the two interpretations is perhaps one of suggestion or connotation only. In an indeterministic universe, some individual events occur for no (sufficient) reason at all. If the law (N:*P*)(*F,G*) is real, and *b* is both *F* and *G*, there could *prima facie* be one of two cases. The first is that *a*'s being *F* necessitated ("brought along with it") its being *G*, in virtue of *F* and *G* and the relation (N:*P*) between them. The second is that *b*'s being *F* is here conjoined with its being *G* as well, but accidentally ("by pure chance"). Now the first reading ("probability of necessitation") suggests that this *prima facie* division may have real examples on both sides of the divide. The second reading suggests that on the contrary, if *F* bears (N:*P*) to *G*, and *b* is both *F* and *G*, then *b*'s being *F* cannot just be conjoined with its being *G* but must have necessitated (to degree *P*) its being *G*. On the first reading there can be cases of something's being *F* and *G* which are not instances of the law, on the second not. But I think either reading could be strained so as to avoid either suggestion. We should therefore consider both possibilities "in the abstract." (In correspondence Professor Armstrong has told me which of the two alternatives he intended, but I shall continue with both, because some readers might otherwise conjecture that the account could be saved by switching to the other one.)

Suppose[2] first that the law (N:*P*)(*F,G*) is real with *P* = 3/4, and that there are three sorts of *F:* those which are not *G,* those whose being *F* necessitated their being *G,* and those which are *G* by pure chance. What is the probability that a given *F* is of the second sort? Well, if *P* is the probability of necessitation, then the correct answer should be *P.* What is the probability that a given *F* is of the third sort? I do not know, but by hypothesis it is not negligible. So the overall probability that a given *F* is a *G,* is non-negligibly greater than 3/4. This is similar to a previous problem, arrived at by a different route, so I won't belabor the point.

Suppose on the other hand that the third sort must be absent (due to some aspect of the meaning of "(N:*P*)"). Then if an *F* is *G,* it is of the second sort. Let us again ask: what is the probability that in the case of a given *F,* its being *F* bears (N:*P*)(*F,G*) to its being *G?* On the supposition that it is a *G,* the answer is 1; on the supposition that it is not *G,* it is 0; but what is it without suppositions? We know what the right answer should be; but what is it? The point is this: by making it analytic that there can be no difference between real and apparent instances of the law, we have relegated (N:*P*)(*F,G*) to a purely explanatory role. It is what makes an *F* a *G* if it is, and whose absence accounts for a given *F* not being a *G* if it is not. (It is not like a propensity, another denizen of the metaphysical deep, which *each* radium atom is supposed to have, giving *each* the same probability of becoming radon within a year.) So we still need to know what is the probability of its presence, and this cannot be deduced from the meaning of "(N:*P*)" any more than God's existence can be deduced from the meaning of "God." It cannot be analytic that the objective probability, that an instance of (N:*P*)(*F,G*) will occur, equals *P.*

Armstrong does consider a different approach, or a different sort of statistical law. Suppose, he says, it is a law that a certain proportion of *F*s are *G*s, at any given time, but individual *F*s which are *G*s "do not differ in any nomically relevant way from the *F*s which are not *G*s." (This is what he could not say in connection with the approaches examined above.) The law would govern a class or aggregate. If the half-life law of radium were construed that way, we would say: if this bit of radium had not decayed, then another bit would have, so that it would still have been the case that exactly half the original radium would remain after 1,600 years. But (as

Richmond Thomason has pointed out in a paper about counterfactuals) we would most definitely not say that. We would say that if this bit of radium has not decayed, then less than half the radium would have decayed. There would be no contradiction with the theory because the half-life of 1,600 years only has an overwhelmingly high probability, not certainty. I suppose there could be another sort of physics in which the half-life law is a deterministic law, and (like in ours) individual radium atoms do not differ "nomically." If Armstrong's account of laws fits that other law of radio-active decay better, that is scant comfort if it cannot fit ours.

5. Laws and Scientific Methodology

It is interesting to compare Reichenbach and Hempel on the famous gold cubes example. Reichenbach, we recall, mentioned "No gold cubes are larger than a cubic mile" as an example of a sentence that is likely to be true, and would apparently pass any formal or general semantic test for expressing a law, but does not. He said that this universal, qualitative statement cannot be accepted as a law by us, because we cannot establish it either by deduction or by induction. The situation would be different for people who had a science implying that such large gold cubes would suffer gravitational collapse. Hempel (1966, p. 55), who took universal generalizations to be confirmed by their instances, said instead that the similar "All bodies consisting of pure gold have a mass of less than 100,000 kilograms" is well confirmed by our evidence—so the reason it is not a law must be something else.

Dretske, Tooley, and Armstrong do not refer to these discussions, but would here side with Reichenbach. They hold that a mere universal generalization cannot, as such, have evidential support. Thus the idea that laws are no more than this would reduce us to skepticism. This "thus" is of course based on a catalogue of alternatives. They do not range far, but base their discussion on a consideration of three philosophical ideas about confirmation, namely *enumerative induction, instance confirmation,* and *inference to the best explanation.*

Dretske puts the main point graphically enough: "laws are the *sort* of thing that can become well established prior to an exhaustive enumeration of the instances to which they apply. . . . [But] it is hard to see how confirmation is possible for universal truths" (p. 256).

To elaborate the first assertion he adds: "Our confidence in them [i.e., laws] increases at a much more rapid rate than does the ratio of favorable examined cases to total number of cases." What exactly does this mean? Not, surely, that if we *know* that something is a law, our confidence increases in this way. Is it rather that if we notice about ourselves that our confidence is increasing in that fashion, we have evidence it is a law? But it could be evidence only of our psychological state—perhaps of an (unconscious?) prior belief that it is a law? Or does it mean that, if the hypothesis is that *it is a law that P,* rather than merely that *P,* our confidence increases more quickly when confronted with the same evidence? Unfortunately, *It is a law that P* implies *P,* so as our confidence in the former increases, confidence in the latter is increasing right along, surely?

To support the second statement, about confirmation of mere regularities, Dretske gives an example of an attempt to confirm a universal statement: *All of the next ten tosses of this coin will come up heads.* Suppose that before we begin, you had selected the coin yourself and had been allowed to test the coin in any way you like, and there was every indication that it was a fair coin. I begin to toss and the first, second, third, . . . toss come up heads. Is your confidence in the hypothesis increasing at all? No. So the positive instances do not appear to confirm the statement, nor are you affected by the observed relative frequency of successes so far.

The ills of enumerative induction have been known since Aristotle—though often wistfully ignored—and those of instance confirmation proportionally almost as long. Many sorts of examples (due to I. J. Good, R. Rosenkrantz, and A. Edidin a.o.) show that positive instances do not in general confirm. The reason is that, typically, they are found in a context in which (a) we have some relevant background information, and (b) the data tells us more than that the items examined are positive instances (see my 1983 for a review). Under these conditions, the positive instances may disconfirm, neither confirm nor disconfirm, or even refute the generalization. Dretske is certainly right that there is something wrong with those philosophical fantasies about the nature of evidential support—but the difference between law and regularity is not it.

In addition there is something curious about how Dretske's position could be established. Suppose I find some example of a statement of universal regularity which is not supported by its instances. Let it be *All Fs are Gs*, and I have (only) positive instances $a_1, \ldots, a_n$, and they do nothing to confirm it. Consider now the hypothesis that it is a law that all *F*s are *G*s. The observed $a_1, \ldots, a_n$ are positive instances of this law. If they confirm the law, I note that the law entails *All Fs are Gs*, which is therefore also confirmed. So my original assertion, that I have found an example of a universal statement which is not confirmed by its positive instances, was wrong. If the reply is that it is clear that it is not a law that all *F*s are *G*s in this case, we have also an example of a law-*hypothesis* which is not confirmed by its instances.[3]

Despite these difficulties, it is not hard to see and appreciate the core idea. It is that a law may be offered in explanation where a mere regularity cannot, and that the goodness of the explanation is evidence (in a way that mere fitting of the data cannot be) for the truth of the hypothesis that there really is such a law. This central use of the idea of Inference to the Best Explanation in epistemology—call it IBE for short—is explored in more radical form by Armstrong.

6. Armstrong on Laws and the Justification of Induction

One conviction expressed by Armstrong at the outset was this: If there are no laws, then the phenomena need not, in any respect, continue as before, and therefore are not predictable. Armstrong notes that this can be denied, a point which he says was brought to his attention by Peter Forrest:

> There is one truly eccentric view. . . . This is the view that, although there are regularities in the world, there are no laws of nature. . . . This Disappearance view of law can nevertheless maintain that inferences to the unobserved are reliable, because although the world is not law governed, it is, by luck or for some other reason, regular. (p. 5)

But he replies immediately that such a view cannot account for the fact that we can have *good reasons to think* that the world is regular.

Armstrong devotes section 4 of his chapter 5 ("The Problem of Induction") to an argument intended to make good on this reply. The truly eccentric view that there are no laws of nature would doom us to inductive skeptism—because the view would leave us without the justification of induction he presents, and because there can be no other. It will be worth our while to analyze this argument in detail.

The argument begins with the explicit premise (call it P_1) that "ordinary inductive inference, ordinary inference from the observed to the unobserved, is . . . a rational form of inference." On questions about what that form is, what rules may be being followed, he confesses himself largely agnostic. He defends the premise along the lines of Moore, common sense against skepticism, saying that this premise is part of the bedrock of our beliefs, indeed, that it "has claims to be our most basic belief of all" (p. 54).

The premise (P_1) is theoretically loaded despite the accompanying agnosticism on questions of form. It is undoubtedly true that we have expectations about the future, and opinions about the unobserved, and that these expectations and opinions change much of the time in a rational manner. It does not follow that we are engaged in ampliation—let alone some sort of ampliative *inference,* i.e., ampliation in accordance with rules—except perhaps at those very points where no reasons can be given (beyond the defense that the move is within the bounds of rationality). In contrast, the strict Bayesian model (surely not to be ignored even if disputable) lets our state of opinion change only by conditionalization on propositions which nature gives us—and conditionalization is merely a logical adjustment. (I use "logical" here very deliberately: what is called deductive logic and the probability calculus are not different except that the former can be used to adjust opinion only by those whose opinions are all of the "certainly, yes" and "certainly, no" sort.) While reliance on our fashion of amending expectations and opinions may be, at least by and large, a basic feature of our psychology, that is not Armstrong's premise. His actual premise, that we are engaged in ampliative inference, which is rational, is a philosophical thesis—recommended as far as I am concerned neither by its origins nor by its subsequent history.

Besides the explicit first premise P_1, therefore, we must count as further premise the statement that we believe the initial premise and that we either know or rationally believe it to be true. This extra premise (call it P_2) is not merely support for the initial premise, but is needed to understand the subsequent argument.

Suppose now that we do engage in a certain form of ampliative inference, which we believe to be rational. At this point in the argument, we need not yet know what that form is (one form that fits all ampliative inference is "P; therefore Q," but the "ordinary inductive inference" presumably includes much less than everything fitting that form)—so let us call this form F.

The next step is a subargument to the effect that it is a necessary truth that induction (i.e., in our present terms, ampliative inference of form F) is rational. This argument is based on P_2 and is an interesting variant on Peirce's argument for the reality of laws of nature ("we know that this stone will drop if released, but how could we know it if . . ." to paraphrase his famous lecture "The Reality of Thirdness"). P_2 says that we know or rationally believe induction to be rational. But that implies that our belief that induction is rational must have a justification. This justification cannot be by induction or it would be circular. Nor can it be deduction, since the relevant statement (i.e., P_1) is not a logical truth and any premise from which P_1 could be deduced would face the same question as we have for P_1, thus leading to a regress. The only possible justification is therefore a claim to knowledge or rational belief not based on any sort of demonstration. That is a tenable claim only for a statement claimed to be known *a priori*. But (I think we must add as further premise) only necessary truths can be known *a priori*. Therefore P_1 is a necessary truth.

I am not sure that to be rational a belief must have a justification reaching back all the way to *a priori* truths. But if we allow, say, *a priori* truths plus the evidence "of my own eyes," as basis for justification, the case for P_1 will not be significantly different. And perhaps Armstrong would go no further, and disagree that belief could be rational unless justified by a demonstration on the basis of *a priori* knowledge plus one's own deliverances of perception. In any case, he has established that if induction is known to be rational, this must be a case of *a priori* knowledge. And if only necessary truths can be known or rationally believed without the sort of justification that P_1 is denied, then P_2 implies that P_1 is a necessary truth.

So now we have, to put it in our terms: it is a necessary truth that ampliative induction of form F is rational. This fact (call it P_3) Armstrong insists, must be given an explanation. What F is will now finally make a difference. He proceeds as follows: he makes a proposal for what F is, demonstrates how P_3 can be true on the basis of this proposal, and then notes that the demonstration would fail if laws said nothing more than mere statements of regularity.

The proposal is that ampliative inference of form F is inference from the evidence, to laws that entail the evidence. He adds that this procedure is an instance of Inference to the Best Explanation (call this P_4), that this sort of inference (IBE) is rational (P_5), and that it is analytic (true by virtue of the meanings of the words) that IBE is rational (P_6). If we add that analytic statements are necessary truths, the explanation of (P_3) is complete. The footnote to be added about the Regularity view of laws is that it would make (P_4) false, because regularities, unlike laws, do not *explain* the evidence which they fit or entail.

The defense of (P_4) must have two parts. The first is that laws explain the phenomena which they fit or entail (I add "fit" to allow, e.g., that a phenomenon could be explained by a law plus some initial conditions, and to allow for probabilistic laws, which may fit, but cannot entail, actual distributions in the real "population"). This first part I think follows from official requirements on the concept of law. The second part is that laws provide the best among the possible explanations that can be given for such phenomena. Let me rephrase this carefully, since one will not generally know initially whether something is indeed a law: evaluating proffered explanations, the claim that the phenomena are thus or so because there is a certain law, will be better than any other sort. This second part of the defense comes in a very cavalier little paragraph:

> It could be still wondered whether an appeal to laws is really the *best* explanation of [the phenomena]. To that we can reply with a challenge "Produce a better, or equally good, explanation." Perhaps the challenge can be met. We simply wait and see. (p. 59)

Would it be enough, to meet this challenge, to present some cases where the best explanation of some phenomena does not consist in deriving them from laws? If so, there is enough literature for Armstrong to confront now; he need not wait.

It will be clear that the defense of the next premise, (P_5)—namely, that IBE is rational—consists in the last premise, (P_6). Armstrong regards it as analytic, as part of the meaning of the word "rational," that IBE is rational. This conviction about IBE appears not only here, but throughout the book, and not surprisingly: IBE is the engine that drives Armstrong's metaphysical enterprise. It provides his view of science (p. 6: "We may make an 'inference to the best explanation' from the predictive success of contemporary scientific theory to the conclusion that such theory mirrors at least some of the laws of nature"). He also regards IBE as being first of all a form of inference to be found pervasively in science and in ordinary life (p. 98): "But I take it that inference to a good, with luck the best, explanation has force *even* in the sphere of metaphysical analysis"—(my italics). To support (P_6) he does not see the need for more than rhetorical questions: "If making such an inference is not rational, what is?" (p. 53); "To infer to the best explanation is part of what it is to be rational. If that is not rational, what is?" (p. 59).

Let me provide an argument which may show the need for more substantial support. Suppose E is our evidence and H is the one and only explanation science has come up with, so that it is the best explanation available. There are many explanations, most of them never formulated, of E even if E is the total evidence of the human race so far. Most of those explanations (perhaps almost all; at least half certainly) are false. I know nothing about H, relevant to its truth-value, except that it explains E. So I must treat H as a random member of this class, and therefore must regard it as likely (perhaps very likely; at least as likely as not) to be false.

To this argument it may be objected that I do know a further relevant fact about H, namely that some person actually proposed H. That is relevant, however, only if we rational animals have a penchant for hitting on the truth—only if actually proposed hypotheses are more likely to be true than others. That is surely not analytic. It may further be objected that the case becomes different if H is the best of the actually proposed hypotheses, and if there were more than one hypothesis; say, twenty-five or twenty-five million. This is still a very small sample, and the premise needed to draw conclusions from it (either that it is a random sample, or that the proportion of truths in it is at least as high as in the whole set) is in any case not analytic either. Only if some support be given for such claims as these will it really be relevant to start interrogating our intuitions about whether better explanations are, ceteris paribus, more likely to be true than worse ones; and *only then,* whether believing one's best available explanation is the rational policy to follow.

Allow me to state a different perspective on this subject. If philosophy is the removal of wonder, as Aristotle said, its proper enterprise is explanation. And producing, elaborating, defending the best explanations you can—providing possible ways of understanding, seeing how things could have been the way they are—is philosophy. Thinking the explanation is true, or that success in explanation is evidence for its truth, is the peculiar characteristic, I think, of the metaphysician—what distinguishes perhaps Plato from the historical Socrates, and Whitehead from Wittgenstein. But to use the model of metaphysics to explain what scientific methodology (and rational adjustment of opinion in everyday life) are like, does not seem to me to have much claim to being even a very good explanation. If Armstrong does not want to defend such a claim, however, despite its central role in his account, what can one say?[4]

7. Whether There Are Laws of Nature

Armstrong characterized the idea that perhaps there are no laws of nature as a "truly eccentric view." After unearthing the assumptions and problems involved in some of its alternatives, that view seems to me no longer so much more eccentric than its opposite. Of course, it has its problems as well. Anyone who held it would have to grant, I think, that we do have the *concept* of a law of nature—and could deny only that this concept has or needs to have a counterpart in reality. For he too would have to account for how we—and he himself—think and speak, in philosophy of science and elsewhere. He would also have to defend views on scientific explanation and methodology of a distinctly, perhaps radically, nominalist/empiricist cast. Even if at all successful in these enterprises, he would still have to face what, at the outset, I called the first motive for taking laws seriously. That is the motive provided by traditional arguments to the effect that there must be, or we must believe there to be, real necessities and laws of nature. This motive remains, and would remain, even if philosophy of science could be entirely demythologized.

So the would-be advocate of that truly eccentric view will, because of the formidable challenge now posed by the more metaphysical writers on laws, have his work cut out. I wish him luck.

Notes

This discussion is part of a longer study in progress [circa 1987]; I would like to thank the National Science Foundation for research support, and David Armstrong, John Collins, Fred Dretske, Richard Foley, Mark Johnston, David Lewis, Wesley Salmon, and Michael Tooley for their kind and helpful discussions and comments.

1. This is not quite correct for Armstrong, who also describes "oaken" laws, the case in which an *F* is *G* provided it does not have some interfering property *H*. He has to consider this a separate case because he does not believe in negative universals.

2. The difficulty raised here was raised previously and independently by John Collins, and discussed in correspondence with Armstrong.

3. I will not discuss Tooley's treatment of these matters here because I understand that he is in process of preparing a new study on this subject: *Causation* (Oxford: Clarendon Press, 1987).

4. Since I am not presenting a rival view here, I run the risk of being thought to be either a strict Bayesian, or a philosophical opportunist willing to use anyone else's objections against his opponent (whether his own view supports those objections or not). So I

would like to add just this: I take "rational" to be a term of permission (what is rational is whatever does not violate standards of rationality, rather than what is rationally compelled). But the idea of an inference has more bite: You cannot rationally accept a conclusion contrary to those which can be inferred from your premises. I think it is often rational to increase the informativeness of one's opinion in a way that goes beyond logical adjustment to the deliverances of experience, but that such moves are not inferences (see further my "Empiricism in the Philosophy of Science").

References

Armstrong, D. M., *What Is a Law of Nature?* Cambridge: Cambridge University Press, 1983.

Dretske, F. I., "Discussion: Reply to Niinuluoto," *Philosophy of Science* 45 (1978), pp. 440–41.

———, "Laws of Nature," *Philosophy of Science* 44 (1977), pp. 248–68.

Earman, J., "Laws of Nature: The Empiricist Challenge," in R. J. Bogdan (ed.) *D. M. Armstrong.* Dordrecht: Reidel Publishing Company, 1984.

Hesse, M., "A Revised Regularity View of Scientific Laws," in D. H. Mellor (ed.) *Science, Belief and Behaviour.* Cambridge: Cambridge University Press, 1980.

Lewis, D., "New Work for a Theory of Universals," *Australasian Journal of Philosophy* 61 (1983), pp. 343–77.

Mellor, D. H., "Necessities and Universals in Natural Laws," in D. H. Mellor (ed.), *Science, Belief and Behaviour.* Cambridge: Cambridge University Press, 1980.

Niinuluoto, I., "Discussion: Dretske on Laws of Nature," *Philosophy of Science* 45 (1978), pp. 431–39.

Peirce, C. S., "The Reality of Thirdness," in C. S. Peirce, *Essays in the Philosophy of Science* (edited by V. Thomas). Indianapolis: Bobbs-Merrill Co., 1957.

Reichenbach, H., *Elements of Symbolic Logic.* New York: Macmillan, 1947.

———, *Nomological Statements and Admissible Operations.* Amsterdam: North-Holland Publishing Company, 1954. (Reissued as *Laws, Modalities, and Counterfactuals.* Berkeley: University of California Press, 1976.)

Tooley, M., "The Nature of Laws," *Canadian Journal of Philosophy* 7 (1977), pp. 667–98.

Van Fraassen, B. C., "Essences and Laws of Nature," in R. Healey (ed.), *Reduction, Time and Reality,* Cambridge University Press, 1981.

———, "Theory Confirmation: Tension and Conflict," in *Epistemology and Philosophy of Science.* Proceedings of the Seventh International Wittgenstein Symposium. Vienna: Hoelder-Pichler-Temsky, 1983.

———, "Empiricism in the Philosophy of Science," in P. M. Churchland and C. A. Hooker (eds.), *Images of Science: Essays on Realism and Empiricism, with a Reply from Bas C. van Fraassen.* Chicago: University of Chicago Press, 1985.

7

Confirmation and Law-likeness

ELLIOTT SOBER

> That a given piece of copper conducts electricity increases the credibility of statements asserting that other pieces of copper conduct electricity, and thus confirms the hypothesis that all copper conducts electricity. But the fact that a given man now in this room is a third son does not increase the credibility of statements asserting that other men now in this room are third sons, and so does not confirm the hypothesis that all men now in this room are third sons. . . . The difference is that in the former case the hypothesis is a *lawlike* statement; while in the latter case, the hypothesis is a merely contingent or accidental generality. Only a statement that is *lawlike*—regardless of its truth or falsity or its scientific importance—is capable of receiving confirmation from an instance of it; accidental statements are not. (Goodman 1965, p. 73)

In this passage from *Fact, Fiction, and Forecast,* Nelson Goodman suggests that a generalization of the form "all *A*'s are *B*" is confirmable by an observed instance (that is, by something observed to be both *A* and *B*) only if the generalization is law-like.[1] Although Goodman has a good deal to say about what makes a generalization law-like, I take it that the basic notion is that law-like generalizations "support counterfactuals"; if "all *A*'s are *B*" is law-like, then if the generalization is true, so is the counterfactual "if something were an *A,* it would also be a B."

Jackson and Pargetter (1980, p. 423), endorsing an objection made by Pap (1958) among others, deny that confirmation requires law-likeness:

> In any strict sense of "law" very few of the generalizations we confirm in everyday life are laws. Surely we have (nonexhaustive) instantial confirmation of "All wines labelled 'Sauternes' are sweet wines," without having to suppose that this is a law of nature. One of Goodman's own examples is the hy-

From *Philosophical Review* 97 (1988): 93–98. Copyright 1988 Cornell University. Reprinted with permission of the publisher and the author.

> pothesis "All men in this room are third sons." . . . He contends that this evidently non-law-like hypothesis is not confirmed by positive instances (short of an exhaustive set including all the men in the room). This is very hard to believe. Suppose the room contains one hundred men and suppose you ask fifty of them whether they are third sons and they reply that they are; surely it would be reasonable to at least increase somewhat your expectation that the next one you ask will also be a third son. Again, surely "All the coins in my left pocket are francs" supports "All the coins in my right pocket are francs." But the relevant universals are notoriously not law-like.

Jackson and Pargetter go on to argue that there is an important insight here, which Goodman's formulation overstates:

> Certain αs being β confirming all αs being β is of interest to the extent that it confirms the universal by confirming certain *other* αs being β. But if the sample of αs are all β *purely by accident,* how could this rationally lead one to suppose a continuation of the connection between α and β outside the sample? We urged that a sample of third sons might (weakly) confirm everyone in the room being a third son, despite the latter's evidently non-law-like nature. But in taking such a sample to so confirm, you take it that its existence is not a fluke, that there is some *reason* for the third sons in the sample being in the room. You cannot take its existence both to confirm and to be a fluke. . . . The important insight then is that one can have no reason to expect the purely accidental to continue. (p. 424)

They then formulate a nomological requirement on the characteristics of a sample, if that sample is to provide instance confirmation of a generalization.

Jackson and Pargetter propose a substitute condition, since they hold that these considerations suffice to discredit Goodman's nomological requirement. However, the initial reaction that Jackson and Pargetter describe is subject to a natural reply by Goodman, one which they anticipate in the passage just quoted. If examining fifty individuals in the room and finding each to be a third son is to confirm the claim that all people in the room are third sons, there must be some reason that the two properties cooccur. For example, it would suffice if admission to the room *caused* an individual to be a third son. Or, more realistically, a non-accidental connection would be assured if people were admitted to the room only if they were third sons. In the absence of a connection of this sort between the two properties, it is hard to see how instances could confirm the generalization.

However, this train of thought makes the generalization start to look law-like. There is now an interpretation of "if *X* had been admitted to the room, *X* would have been a third son" on which this counterfactual comes out true. It isn't that admission to the room would have turned an only daughter into a third son. Rather, the idea is that if *X* had been admitted, *X* would have had to have been a third son.

A similar story can be concocted for the coin case. If the contents of my left pocket are to count as evidence for what is in my right, then there must be something about the processes by which those two pockets were filled that licenses certain counterfactuals. Once those processes are acknowledged, the relevant generalizations start to sound law-like. It will now be true that if *X* had found its way into either pocket, *X* would have had to have been a franc. A similar line of reasoning presents itself for the generalization about Sauternes.

This reply provides no easy victory for Goodman, in that it is no determinate matter whether the required counterfactuals are true. If the mechanism of admitting only third sons is in effect, what becomes of the counterfactual "if *X* were admitted, *X* would be a third son"? One can argue that it is true and appeal to the mechanism. Or one can argue that it is false and say that the admission of certain individuals would have required the mechanism to break down. Some may see here an objective question. I see a pragmatic one in which both readings have their place. This strategy of argument gets Goodman out of trouble, not by turning defeat into victory, but by achieving a stalemate.

However, the objection that accidental generalizations can be confirmed by their instances is sustainable against this reply. One can *know* that a generalization is purely accidental and still have positive instances count as confirming. Imagine an urn that is filled with a thousand balls by drawing from a source whose composition is known. Suppose the source contains 50% red balls and 50% green balls. By random sampling, the urn is composed. The inference problem is to sample (with replacement) from the urn and infer what percentage of red and green balls it contains.

Since we know that the urn was composed by draws from the source, we can assign prior probabilities to each of the possible compositions, from 1,000 red and 0 green to 0 red and 1,000 green. When we sample from the urn, we can use Bayes's theorem to compute the posterior probability of the various hypotheses about the urn's composition. Suppose I sample 250 times from the urn and find that each sampled ball is green. These observations make the hypothesis that all the balls in the urn are green more probable than it was initially. I thus have obtained instance confirmation of the generalization. But knowledge of the process whereby the contents of the urn were assembled assures me that this generalization, if true, is only accidentally so. It is a mere fluke if all the balls in the urn happen to be green. There is nothing about being a ball in this urn that makes something green, nor is it true that a ball would not have been put into the urn unless it were green.

It makes all the difference whether the mechanism whereby the population is assembled is known in advance or is inferred in the process of sampling. When fifty individuals in the room are sampled and are found to be third sons, the suspicion naturally arises that this is not an accident. The same may be said when the balls sampled from an urn are all found to be green. However, the fact that nomological connections are reasonably suspected in such cases does not show that law-likeness is a presupposition of instance confirmation. Simply replace prior ignorance of process with a substantive process assumption (of the kind just sketched for the urn problem) and the composition of a population known to be fortuitously assembled can be confirmed by random sampling.

In the passage quoted at the beginning of this article, Goodman enunciates a second thesis about confirmation. Besides connecting confirmation and law-likeness, he also claims that if observed instances confirm a generalization, they also must confirm claims about unobserved instances. A small modification in the inference problem just sketched will now show that these two confirmational relationships sometimes part ways.

Let us proceed exactly as before, except that we now sample without replacement. In this case, the probability that the next ball will be green is 0.5, regardless of what the previously sampled balls were like. In spite of this, a generalization about the whole urn can be confirmed. Observing 250 green balls confirms the hypothesis that all the balls in the urn are green by raising its probability, but this observation does not raise the probability that the next ball drawn from the urn will be green. The simple case in which the urn contains just two balls illustrates the general point. When the balls are drawn without replacement, there are four equiprobable sequences of green (*G*) and red (*R*) draws—*GG, GR, RG,* and *RR*. If the first ball sampled is green, then the probability of *GG* increases from 0.25 to 0.5. However, the probability that the second ball will be green is 0.5, just as it was before any ball was drawn. I conclude that confirming a generalization and confirming a claim about the next instance are not always as intimately connected as Goodman suggests.

It remains to show that Jackson and Pargetter's substitute nomological condition fares no better than Goodman's. Suppose I sample from objects that are all Y_1 and use my observations to infer what the objects in Y_2 are like. Jackson and Pargetter (p. 424) require that "All Y_1 *A*'s are *B*" confirms "All Y_2 *A*'s are *B*" only if the objects sampled would still have been both *A* and *B* even if they had been Y_2 instead of Y_1. It is worth noting that this condition is not really an explication of the intuition that Jackson and Pargetter earlier noted—namely, that a sample cannot confirm if its existence is a fluke. In the case of random draws from a heterogeneous urn, the composition of a sample may be said to be a "fluke" if it is improbable enough, even though the objects sampled would have had the same characteristics had they not been sampled. The case to be noted next exhibits the nonequivalence of these two conditions in the other direction: the sample is no fluke, but the objects sampled do not have the nomological property that Jackson and Pargetter demand. The example now to be constructed also shows that their nomological condition is not necessary for instance confirmation.

Warm-blooded organisms regulate their temperature physiologically; cold-blooded creatures do not, but achieve the same result by moving from one microenvironment to another. So, for example, human beings perspire when placed in the hot sun, whereas lizards simply crawl into the shade. Suppose I believe that human beings and lizards have the same body temperature, whose exact value I wish to confirm by observation. I then measure some lizards in their normal environments and find that all of them have a body temperature of 98.6°F. My background beliefs lead me to expect that human beings will have the same body temperature. However, I then note that the lizards were all measured in the shade, whereas some human beings spend a great deal of time in the blazing sun. This difference does not confound my inference, even though Jackson and Pargetter's condition goes unsatisfied; it is false that the lizards would have exhibited the same temperature if they had been measured in the sun. "All organisms living in the shade have a temperature of 98.6°" can confirm "All organisms living in the sun have a temperature of 98.6°," even if the shaded organisms would not have had that temperature if they had lived in the sun.

I very much doubt that fine-tuning Goodman's proposal or the one Jackson and Pargetter suggest will improve matters. Observations have confirmatory significance only within the context of a set of background assumptions (Good 1967; Rosenkrantz 1977). Stipulate a nomological condition on instance confirmation and a background context can be invented that shows that condition to be unnecessary. This is not to say that matters of modality are irrelevant to questions about confirmation. Rather, what I doubt is that there is a single "nomological condition" that each and every act of confirmation must obey. The multiplicity of possible background assumptions means that there will be many ways that observed instances may confirm generalizations (and many ways in which they may fail to do so). The nomological conditions just discussed should be viewed as parts of such *sufficient* conditions. But this is a far cry from identifying the nomological assumptions underlying every case of instance confirmation.

Notes

1. The view is also endorsed by Scheffler (1965) and Strawson (1952).

References

Good, I. J. (1967). "The White Shoe Is a Red Herring," *British Journal for the Philosophy of Science* 17, p. 322.

Goodman, N. (1965). *Fact, Fiction, and Forecast.* Indianapolis, Ind., Bobbs-Merrill.

Jackson, F. and Pargetter, R. (1980). "Confirmation and the Nomological," *Canadian Journal of Philosophy* 10, pp. 415–28.

Pap, A. (1958). "Disposition Concepts and Extensional Logic," *Minnesota Studies in Philosophy of Science,* Vol. II, edited by H. Feigl et al. Minneapolis, Minn., University of Minnesota Press.

Rosenkrantz, R. (1977). *Inference, Method, and Decision.* North Holland, D. Reidel.

Scheffler, I. (1965). *The Anatomy of Inquiry.* Indianapolis, Ind., Bobbs-Merrill.

Strawson, P. (1952). *An Introduction to Logical Theory.* London, England: Methuen.

8

The World as One of a Kind

Natural Necessity and Laws of Nature

JOHN BIGELOW, BRIAN ELLIS, AND CAROLINE LIERSE

1. Introduction

This world is one of a kind. Some philosophers have maintained that there are many other worlds which are spatially, temporally, and causally unrelated to ours. We are not asserting that there are any such disconnected worlds. Nor do we assert that there are none. There is at least one world; and it is a member of a natural kind whether or not there are any others of its kind. If there were any other world, in addition to this one, or instead of this one, then there would be a nontrivial question whether that world was of the same natural kind as ours. We can imagine worlds which would be of the same natural kind as ours; but we can also imagine worlds which would not.

Recognition that this world is one of a kind offers a new approach to the question of what a law of nature is. We argue that in general laws of nature are concerned with natural kinds. In some cases laws simply describe the essential properties of natural kinds. Maxwell's equations, for example, describe the essential properties of the electromagnetic field. In the case of other laws, for instance where there are interactions between things of different kinds, the laws stating how they behave are derivable from their essential natures. We hold that this is true even of the most fundamental laws of nature, e.g., the conservation laws, the principles of relativity, and the symmetry principles: they too are concerned with the essential properties of a natural kind. Their concern is with the kind of world this is.

This theory of the nature of scientific laws derives from the basic idea that things behave as they do because of what they are made of, how they are made, and what their circumstances are. Insofar as the behavior of a

From *British Journal for the Philosophy of Science* 43, 3 (1992): 371–88 by permission of Oxford University Press.

thing depends on what it is made of, it depends on the essential natures of its constituents. Insofar as its behavior depends on how it is made, or on its circumstances, it depends on laws of interaction. The various laws of interaction, we suppose, depend on the essential natures of the things interacting with each other, and on the forces or fields which mediate these interactions. Thus we hold that laws of nature are grounded in the essential natures of things, not superimposed on them, as the term "law of nature" seems to suggest.[1]

2. Natural Kinds and Essences

When we speak of natural kinds, we have in mind things like copper, gold, protons, or electromagnetic fields. They are kinds of things which exist in the world independently of human knowledge, language, and understanding.

Not every class of things counts as a natural kind. For every objective property there is a class of things which have that property. But not every such class constitutes a natural kind.

Natural kinds of the sort we have in mind are characterized by *clusters* of properties which play an especially important *explanatory role* with relation to other properties and relations. They are what Mill called "real kinds," and are to be contrasted with classes of things which just happen to share some number of properties, but which do not share any cluster of *explanatorily crucial* properties.

If we are to use natural kinds in the explication of laws of nature, then we must sooner or later give an analysis of natural kinds. We presuppose that some decent analysis of natural kinds can be given. We presuppose, furthermore, that natural kinds may be taken as *more basic,* in some sense, than laws of nature. If we are to use natural kinds to explain *laws,* then we should not use laws to explain *natural kinds.* To do so would be to move in a tight circle.

There are at least two ways of proceeding to construct such an explanation. The first is to develop an ontology in which natural kinds occur as primitives. For example, we might suppose natural kinds to be primitive substances which are distinguished from each other by their internal properties and structures, and whose dispositions to act or react are determined by these internal properties and structures. These substances may be supposed to exist no less primitively than the things which instantiate them.

The essential properties of a natural kind might then be identified with those properties which belong strictly to the natural kind itself, while its so-called accidental properties might be supposed to be just those properties which happen to be instantiated by all or some of the individuals which are its instances. In this way, we may be able to provide an ontological foundation for the essence-accident distinction.

The second approach is to develop a theory of essences as primitive, and seek to define natural kinds in terms of essences. Essential properties can be defined in terms of necessity, quantification, and identity. An essential property of a given individual is one which is such that, *necessarily,* if *anything* lacks that property, then that thing is not numerically *identical* with the given individual. Essential properties, thus defined, could then be used to define natural kinds. Membership in a natural kind is, arguably, an essential property for each of the members of that kind. Consequently natural kinds could be construed, for instance, as classes of individuals which share an explanatorily significant cluster of essential properties.

However, it is beyond the scope of this paper to develop a theory of natural kinds sufficient to support the theory of laws which is being proposed. Here we merely aim to show what consequences would follow concerning laws of nature, given that we are right about the ontological priority of natural kinds and essences over laws of nature.

If this world is granted to be one of a natural kind, then natural necessities will emerge in the following way. Natural kinds are always associated with *essential properties.* If something is of a natural kind, then there will be properties which this thing must have to be a thing of that kind, and which it could not cease to have without ceasing to be a thing of that kind. We think this applies to the actual world: there are properties which this world could not lack without ceasing to be the kind of world it is. Other properties of this world are *accidental properties.* A world of the same kind as ours could exist which lacked these properties. Worlds which lacked the essential properties of our kind of world could also exist. They are not *logically* impossible. But they would be different kinds of worlds from ours, and so not other ways *this* world could have been.

Laws of nature, we claim, derive from the attribution of essential properties to things. A special case will be that of the attribution of an essential property to the actual world. Such an attribution will be a posteriori, since we do not know a priori which possible world we are in. Yet if it is a correct attribution, it will not be on the same footing as a sheer contingent truth about the actual world. It will have a modal status intermediate between sheer contingency on the one hand, and logical necessity on the other. It is not like logical necessities, which are true of any world whatever. Yet it is not like a sheer contingency, which could have either held or not held of the world we are in. It has a degree of necessity, which we may aptly call *natural* necessity, since it depends on the nature of the world which we happen to be in. A law of nature is not just something which is true of the actual world but which could have been otherwise *in this very world;* rather, it is something which could not have failed to hold *of this world* without this world ceasing to be, and another world altogether existing instead—another world of a different natural kind with a different nature from the one we are in. Thus, there is an immediate advantage to a theory which construes laws as attributions of essential properties: it automatically bestows upon laws the kind of necessity which they manifestly do have, something intermediate between mere contingency and logical necessity.

On traditional Humean conceptions laws are privileged correlations. Yet there are many laws, we shall argue, which are not easily construed as asserting correlation of one thing with another. Various laws, such as the fundamental conservation laws, symmetry laws, and a variety of other laws must be laboriously kneaded into shape before they can be fitted into the traditional Humean conception of laws as privileged correlations. On our theory no such kneading is needed: laws may often be correlational, but they need not be, as we shall show in Section 4.

According to Aristotle, the essential properties of a thing are those on which its identity depends. They are the properties of a thing which it could never have lacked, and which it could not lose without ceasing to exist or to be what it is. Accidental properties of a thing are properties which that thing could either have had or lacked. For example, you could be sick, or you could be healthy, so your state of health is not one of your essential properties. An essential property of you, if you have any, would be a property which you could not lack without ceasing to be who and what you are. If Aristotle was right, then being human is one of your essential properties, for the property of being human is not something you could lose without thereby ceasing to exist, i.e. without ceasing to be who and what you are.

However, when we speak of laws as attributions of essential properties we are not thinking of such properties in the Aristotelian sense; that is, we do not construe essential properties just in terms of the conditions deemed necessary and sufficient for the application of some particular concept.

Rather, what we have in mind are *real* essences in Locke's sense. According to the Lockean conception, real essences are properties which play a central role in the scientific explanation of a thing's other properties and relations—they are the characteristics which make a thing the kind of thing it is. For example, consider the case of water being essentially H_2O. According to the Lockean conception, being H_2O is an essential property of water, for its being H_2O is what makes water the kind of substance it is. Nothing could come to have the structure H_2O without water thereby coming into existence. Likewise, some water could not cease to have the structure H_2O without that water ceasing to exist.

It is worth noting that our paradigms for essences come from physics and chemistry rather than, as for Aristotle, from biology. Essences are easier to grasp for relatively simple substances like water; they are harder to grasp for complex, self-replicating, and evolving substances like cows or grass. Ultimately, the causally explanatory Lockean essence for an organism could arguably be taken as its genotype, and this is a chemical structure like H_2O, only much more complicated. Yet the genotype for Socrates is not strictly identical with that of Callias or of anyone else. So genotype seems not to be quite the same as the supposed Aristotelian essence of humanity. Aristotle held that all humans have exactly the same essence. One way of understanding that idea is by comparing it with the now unacceptable hypothesis that all humans have exactly the same genotype, and that all differentiation within the species is phenotypic. What has been discovered in biology, however, is that all that holds a species together as a single natural kind is a sufficient *degree of similarity* among genotypes.

When we turn our attention to something even more complex than cows and grass, namely to *the whole world,* essences and natural kinds will probably be even more a matter of degree than they are for cows and grass. But this will not empty the idea of all content. Just as for each cow there will be its own fully determinate individual bovine genotype, so too for the world we inhabit there will be its own fully determinate set of essential properties—the complete catalogue of all its natural laws. Just as a cow's genetically identical twin, if it has one, will have exactly the same Lockean essence, so too another possible world with all the very same laws as ours will be unambiguously another way this very world could have been—a physically possible world with respect to our world. As we turn our attention to worlds whose laws resemble ours less and less completely, we shade off by degrees into worlds which are less and less of a kind with our world. Eventually we reach worlds which are as undeniably of a different kind from our world as cows are of a different biological kind from the grass they eat.

Thus, the Lockean conception of essences and natural kinds can be applied, without absurdity, to complex as well as to simple individuals. There is no reason in principle why it could not be applied to as big an individual as the whole world. The Lockean real essence of the world itself would have to consist of the properties and structures which make this world the kind of world it is. They would have to be those which have a fundamental role in explaining the patterns of events observed in the world.

The thesis to be defended here is that the fundamental laws of nature derive from the Lockean essential properties and structures of the world. If our analysis is sound, then laws of nature must be supposed to exist, and to hold *necessarily,* whether or not anybody knows that they exist, or has any idea of what they are. Hence, our theory of laws is an *ontological* one. It grounds laws in something which is postulated to exist independently of human expectations or conventions. Not all philosophers agree that laws need any such ontological grounding. Humean theories explain why laws *seem* to be necessary even though there is nothing in the world which grounds this necessity. Some philosophers have tried to explain why it is desirable to hold on to some generalizations more firmly than others, making them effectively true by convention, thus giving them the status of provisional conceptual truths. Theories of this sort make no postulation of any *ontological* grounds for laws.

However, in this paper, we make no attempt to refute any of the non-ontological theories. Nevertheless, it is important to contrast our theory of laws of nature with other ontological theories which purport to explain their nomic necessity, and to show that our theory has many advantages over them.

3. The Direction of Explanation

The best-known and best-developed of the current ontological theories are those of F. I. Dretske, M. Tooley, and D. M. Armstrong (DTA).[2] While these three accounts are distinct, they are sufficiently similar for us to be able to treat them together.[3] For the purposes of this paper we shall focus mainly on Armstrong's theory of laws.

Armstrong claims that laws of nature are irreducible dyadic relations of necessitation (or probabilification) holding between universals. It is this relation between two universals (itself a universal) which is supposed to endow certain regularities with their nomic status.

Although Armstrong's approach has some plausibility, we do not think that this account of nomic necessity is successful. Nor do we believe that this theory of laws is tenable in the light of current scientific theory. But our purpose here is not to offer a refutation,[4] but rather to propose an alternative ontological theory of the nature and necessity of laws, and to clarify our theory by contrasting it with its current rivals.

The most striking difference between our account of the necessity of laws and the accounts offered by DTA concerns the *direction of explanation.* On our account, laws express, or are derivable from statements expressing, the essential nature of the world and the kinds of things it contains. The direction of explanation is from the bottom up, so to speak, *from things* with their necessary properties *to laws.*[5] On the DTA account, laws are viewed as correlations between the properties or behavior patterns of various kinds of things in the world, and their necessity is explained by the existence of certain "necessary" second-order relations between the universals that are correlated. Thus, the direction of explanation is from the top down, so to speak, from the second-order relations to the first-order correlations.

While Armstrong's account circumvents some of the more damaging objections to the traditional reductionalist accounts, we argue that its main weakness lies in the difficulty it has in explaining the necessity of the relations between universals. Laws of nature, Armstrong states, are *contingent* relations of necessitation. That is, even when such a relation does tie one universal to another, it is not *logically* necessary that this relation should hold between these two universals—the holding of this relation is a contingent matter.

At first blush the contingent status of Armstrong's necessitation relation seems to have definite merit. Not only does it guarantee the a posteriori character of laws, it also fits well with our belief that such relations (laws) could have been otherwise. But why should it be that some sets of relations are more special than others, thereby constituting "necessitation" relations between universals? One answer requires us to take this relation of necessitation as primitive, and to say that it is just a basic fact that certain relations are necessitation relations. This is Armstrong's position. To illustrate, consider the law that all metals are electrical conductors. According to Armstrong's theory, the necessity of this law derives from the fact that there is a relation of nomic necessitation holding between the properties of being a metal and that of being an electrical conductor, so that anything which has the first property must necessarily have the second. But what makes the relation between these properties one of nomic necessitation? According to Armstrong's theory, nothing does; its status as a necessitation relation is primitive and unanalyzable. It is simply one of a class of necessitation relations which are to be found in nature.

However, this response fails to provide an illuminating explanation of the nomic necessity of laws of nature. On this model, laws or law statements are just *descriptions* of "necessary" relations between universals, but why these relations hold is left unanswered, and, even granted that they do hold, the question remains why the holding of these relations is appropriately described as constituting a "necessitation" relation. Given that the holding of the relation is contingent, what does its holding between two universals have to do with "necessity"? As Lewis points out, *labeling* the relevant relation a "necessitation relation" cannot by itself create a *necessary* link between the related things, any more than calling someone "Armstrong" can give them mighty biceps![6]

One alternative approach to the explanation of nomic necessity is by appeal to higher-order universals.[7] The alleged "necessitation" relation R_1, which purportedly joins properties p and q, must hold of p and q in virtue of something. Suppose it holds in virtue of some further relation R_2 in which R_1 stands to p and q. We might then attempt to explain the "necessity" in the relation R_1 by claiming that it holds in virtue of R_2. However, this line of reasoning immediately threatens a regress. We suppose <p, q> to stand in relation R_1, and we explain that this is a necessitation relation because it involves a distinctive relation R_2. We may then ask why R_2 is distinctive in a way which warrants the title "necessitation." It is, of course, possible to terminate the regress at any time, but this can only be achieved by claiming that the necessity of a particular relation R_n must be taken as primitive. Hence, this approach fails to give an enlightening exposition of the nature of laws.

Like Armstrong, Dretske, and Tooley, we believe that certain relations between universals are in some sense "special." We agree with them, for example, that there is a relation of necessitation holding between the properties of being a metal and being an electrical conductor. But on our view, this relation derives from the essential nature of metals, as well as the essential natures of their elementary constituents and the electromagnetic fields which may act on them. That is, we think that the relation of necessitation between these universals is ontologically grounded in the essential natures of the various kind of things involved.

Laws often entail the presence of relations between universals. These relations are special because they are *necessary* relations. Yet unlike Armstrong et al. we do not believe that their necessity stems from an intrinsic (or primitive) feature of the relation itself. Rather, it is to be explained at a more fundamental level. Laws of nature, we argue, are truths whose necessity is grounded in the essential properties of this world and the things in it. Hence, it is not the relation between universals that constitutes the necessity of laws, rather, their necessity results from the essential natures of the properties on which the nomological relation supervenes. Thus, laws of nature are not just mysterious relations that preside over an unsuspecting world, but relations which bear an essential connection to the world and its content, in that their nature, existence, and necessity derives from the essences of the things they govern.

How is it possible for the laws of nature to arise from essential properties? This can be explained by direct appeal to the nature of a property itself. We argue that included among the essential properties *of a property* is the *propensity* or disposition of whatever possesses it to display a particular kind of behavior in a specific kind of context. What science observes and codifies are the manifestations of these dispositions. Hence (statements of) laws which describe how properties behave will at the same time tell us what things which have these properties *must* do, in virtue of being the kinds of things they are. The necessity of laws is not a primordial feature of necessitation between properties, but rather is "inherited" from the underlying essences of those properties. The necessitation relation between properties is, we argue, *supervenient* on the essences of the properties which stand in that relation.

One important feature to be made clear is that the kind of supervenience we are postulating of laws on essences, is *broad* supervenience:[8] i.e., given a particular set of base properties (the subvenient class), the ensuing supervenient properties or relations are the same in all possible worlds which feature the subvenient properties. This means that, given the essential natures of the kinds of things said to be nomically connected, the supervenient relation (law) must *necessarily* follow. Hence, there is no possible world where things of these kinds exist, but the supervening (nomic) relation fails to hold.

It has been objected that our analysis of natural necessity makes laws logically necessary where clearly they are not.[9] This objection stems from the fact that we explain the necessity of laws in terms of the essential properties of natural kinds. And yet it is not logically possible for a natural kind to have different essential properties from those it does in fact possess—if one of its properties could be absent, it would not be an essential property. Hence, anything which is true in virtue of the essential properties of a natural kind will have to be true in all possible circumstances in which that natural kind exists.

Nevertheless, this fact does not imply that the laws of nature are logically necessary. For it is a contingent matter what natural kinds there are. There was a time when chlorine was thought to have atomic weight 35.5 approximately, and at that time this would have been thought to be an essential property of chlorine. When isotypes were discovered it was found that nothing had atomic weight 35.5; but it was not concluded by anyone that chlorine did not exist. This shows the revisability of our statements about essential properties—they are conjectures about Lockean real essences rather than stipulative definitions of concepts. Hence laws are *epistemically* contingent. This accounts for the a posteriori character of laws. It is consistent with this, however, that laws are true in virtue of the essential properties of natural kinds. *Given* that the natural kinds referred to in the statements of a law are what they are, *then* the truth of the law is determined by the essential properties of those natural kinds. Fix the natural kinds referred to, and you determine the truth value of the statement of the law. So a natural law, unlike a logical truth, would not be true *in all worlds,* but would be true in all worlds which contain the natural kinds mentioned in the law. Hence, a law possesses a kind of conditional necessity: necessity relative to the natural kinds to which it actually refers.

To highlight an important contrast between the rival ontological theories and ours, imagine the world, and the laws within it, as represented by the game of chess. In this analogy, take the board to represent the basic structure of the universe (including its space-time structure, its symmetries, and the conservation laws), the pieces to represent physical entities, and the rules of the game to represent the laws governing the relations between these entities. One salient feature of this picture is that there does not appear to be any *logical* connection between the matter in the world and the laws that govern its behavior. It is possible that the rules for chess could have been formulated so that the legal moves for the rook, for instance, required that it moved only in a diagonal direction. The fact that it is the bishop which moves in this particular way is just a contingent fact about the game—it is *logically* possible that the rules could have been otherwise. In fact, if the rules are imposed on the pieces from above (as by irreducible, necessary relations), it can be seen that the way the rook moves has nothing to do with its essential nature. It is not difficult to imagine dozens of different ways the rule book for chess could have been written, or alternatively, the possibility of the pieces and the board existing without a rule book to govern the movements of the pieces (i.e., an anarchic, Hume world). This image portrays the creator of the game as carving the pieces, forging the board, and then sitting down to formulate the legal movements for each piece. This is like an image of God calling forth the subatomic zoo, letting there be a space-time to house it, and then sitting down to formulate the laws governing their interactions.

In contrast with the DTA account, we urge that there is no need for a chess manual to dictate the rules of movement for the chess pieces. There is no need for a rule book because there is no freedom to choose the rules of movement for the pieces. By embedding laws in essential natures, the legal moves for a given piece are determined by the nature of the piece itself. Thus, by construing rules (laws) as consequences of essential properties, the rules were created simultaneously with, and exist in virtue of, the pieces which they govern.

According to the analogy, the bishop can only move (and thus instantiate a rule) in virtue of its essential nature. The way it interacts with the board and other pieces is what makes it a bishop. If we were to modify the game in such a way that the piece which we previously referred to as the "bishop" now moved according to a different set of rules, then the piece would no longer be a bishop. Hence, by changing the rules for moving a particular piece, we are changing its essential nature, and so changing its identity. For this reason, we argue, it makes no sense to talk about the other ways *the bishop* could have moved.

Consequently, it makes no sense to talk about a piece independently of the rules which govern its moves. Analogously, it makes no sense to speak of a natural kind, e.g., being an electron, independently of the laws which govern its behavior. An electron is an electron precisely because it exhibits a specific set of interactions under various conditions: this is what makes it an electron. If something failed to exhibit these qualities it would not be an electron; there would exist, instead, something other than an electron. Therefore, to speak of a possible world with a different set of laws to ours necessarily entails speaking of a world containing different natural kinds. Hence, "Disney" worlds, in which things defy our law of gravity and so forth, are by logical necessity ruled impossible as ways in which the things in *this* world could have behaved. And if we did, by chance, happen to come across a world which appeared to resemble ours in all respects except for the laws governing the behavior of its inhabitants, its similarity to ours would be illusory. Prima facie, it may appear to contain entities belonging to the same natural kinds as ours, but it could not in fact be an authentic copy of our world.

Thus, our theory of laws differs from its main ontological rival in the following way. In the DTA theory, laws rest on necessitation relations between universals; and the holding of these relations between universals is a *contingent* matter. Properties which stand in a necessitation relation *could have* failed to stand in that relation, *without* thereby ceasing to be the properties that they are. According to our theory, in contrast, such necessitation relations are grounded in the essential properties of the natural kinds which stand in these relations.

It remains to be shown that our conception of laws of nature is a viable one, in that it stands up well as an account of the currently accepted laws of nature. This is the task to which we now turn.

4. Essences and Laws of Nature

Laws of nature all have the kind of necessity which we call "natural necessity." For every action, there *must* be an equal and opposite reaction. If the potential difference along a wire is V, and its electrical resistance is R, then a current $I = V/R$ *must* flow along the wire in the direction of decreasing potential. For a black body radiator, the radiation *must* have a frequency distribution in accordance with Planck's radiation law. In any closed and isolated system total momentum *would have to be* conserved. No two electrons *can* be in the same mechanical state. It is *impossible* for there to be a perpetual motion machine. It is *impossible,* by any finite process, to reduce the temperature of an object to absolute zero. It is *impossible* to accelerate an object of non-zero mass to the speed of light. The question is, "What is the source of this kind of necessity?"

The simplest examples of this kind of necessity are the "laws" of chemistry which concern the basic properties of the various elements and compounds, their atomic and molecular structures, the kinds of bonding which are to be found in them, how they may be ionized, what kinds of salts may be formed from them, and so on. For example, it is essential to the nature of copper that it is a metal with atomic number 29 and electron structure 2.8.18.1; that it forms two series of salts, cuprous and cupric, the cuprous ions being monovalent and having unit charge +1, the cupric ions being bivalent and having charge +2. Similarly, it is essential to the nature of common salt that it is sodium chloride, its molecular formula is NaCl, the bonding is electrovalent; and it is essential to the nature of methane that it has the molecular formula CH_4, molecular structure:

```
      H
      ..
  H : C : H
      ..
      H
```

and that the bonding in this structure is covalent. Clearly, chemistry textbooks are full of "laws" like these.

But are these really laws of nature? Armstrong, for example, claims that these "composition laws," as they are sometimes called, are not genuine laws of nature, because they fail the test of relating distinct, non-overlapping universals.[10] Moreover, they are seldom called "laws" by textbook writers, and most scientists would probably agree with Armstrong that they are just some of the basic facts of chemistry. The laws of chemistry, they are likely to say, are the principles governing chemical interactions, and not descriptions of the fundamental properties of elements and compounds. From our point of view, it does not matter whether we call these descriptions "laws" or not. They are all necessary statements attributing essential properties to natural kinds, and clearly they are an important part of the theoretical basis for determining what chemical reactions are possible or impossible, or in what circumstances they will occur. Therefore, even if they are not themselves laws, it is plausible to suppose that the necessity of the laws governing chemical reactions derives, at least in part, from the essential natures of the chemical elements and compounds involved in these reactions.

Consider, for example, the law that when hydrochloric acid is electrolyzed, hydrogen is liberated at the cathode and chlorine at the anode. The necessity of this law derives from the nature of the bonding between hydrogen and chlorine in hydrochloric acid, the fact that hydrogen ions are positively charged, while those of chlorine are negatively charged, and the fact that opposite charges attract each other.

In particle physics, the equivalents of the chemical elements are the various fundamental particles. Like the elements, they are natural kinds, and we are now able to say quite a lot about their essential natures. For example, the electron is a stable lepton, with unit negative charge and spin 1/2; the proton is a stable baryon, with unit positive charge, spin 1/2; and so on. If you wish to say that these are not laws of nature, just basic facts about the fundamental particles, we do not mind. You may prefer to reserve the title "law" for the principles which determine what particle interactions are possible or impossible, or in what circumstances they may occur. If so, we are willing to go along with this suggestion, for these principles would appear to have a similar status to the laws of chemical interactions, except that they are non-classical probabilistic laws. The same point may therefore be made with respect to the principles of particle interaction: we should expect the detailed explanations of these laws to depend on the essential properties of the fundamental particles. It is plausible, therefore, to suppose that the laws governing these interactions have to hold because it is essential to the natures of the particles involved that they should have the dispositions to interact with each other in the kinds of ways they do.

What goes for particles also goes for fields. Consider Maxwell's equations. It is our view that Maxwell's equations describe the essential properties of the electromagnetic field. To simplify the case, consider Maxwell's equations for otherwise empty space (i.e., where there are no charges or currents). These equations tell us: (a) how the rate of change of the electric field depends on the gradients in the magnetic field; (b) how the rate of change of the magnetic field depends on the gradients in the electric field; and (c) that neither the magnetic nor the electric field is divergent in these circumstances (i.e., there are always as many lines of force entering a region as leaving it).[11]

It is our contention that these equations describe the essential properties of the electromagnetic field. Such a field is necessarily a field which has electric and magnetic components which are interdependent in these ways. If a region of space contains no electric or magnetic field components which are so related, then it does not contain an electromagnetic field.[12] We hold that these equations represent the essential nature of the electromagnetic field for two reasons. First, the equations are evidently quite fundamental to the whole theory of electromagnetism. A very wide range of electromagnetic phenomena may be represented as solutions to these equations. Second, there is no rival conception of the essence of the electromagnetic field which would allow us to view the field equations as anything other than essential. If they are not essential, what is the nature of the entity which bears the varying properties of electric and magnetic field strength which happen to be related by these equations? The ether? No. There is no ether, there is nothing but the field.

If we are right about Maxwell's equations, then this supports an analogous way of regarding other kinds of field equations in fundamental physics. If Maxwell's equations describe the essence of the electromagnetic field, the equations of quantum field theory might plausibly be supposed to describe the essential natures of the fields which are or carry material particles. Thus the scientific endeavor to discover the essences of things is not clearly distinguishable from the quest for the most basic laws of nature. For at least some principles which nearly everyone would describe as "laws of nature," e.g., Maxwell's equations, are directly concerned with essences. The question is whether all genuine laws of nature are descriptive of the essences of natural kinds in the same sort of way, or are derivable from laws which are. We think they probably are, and that this is the explanation of their distinctive kind of necessity.

However, to explain the necessity of the most fundamental laws of nature, e.g., the conservation laws, the principles of relativity, and the symmetry principles, it does not ring true to suppose that they are characterizing one specific kind of thing within the world. On the face of it, such laws neither ascribe properties to things within the world, nor describe correlations between things in the world. It is natural to construe them, rather, as characterizing not natural kinds *within* the world, but the world *as a whole*—as describing the kind of world in which we live.

Consider the conservation laws. These laws say what sorts of events and processes are possible. Basically, what these laws tell us is that in all self-contained events and processes certain quantities must be conserved. The quantities that must be conserved include energy, momentum, angular momentum, charge, lepton number, baryon number, and one or two other quantities. The conservation laws apply directly only to events and processes occurring in closed and isolated systems, i.e., systems where there is no inflow or outflow of matter or energy, and which are not being affected by any external forces. But all systems, other than the universe itself, are acted upon by external forces, e.g., gravitation forces; and, in all macroscopic systems, except for the universe as a whole, there are inflows and outflows of matter or energy. So, strictly speaking, the universe is the only system which is perfectly closed and isolated. However, the external gravitational forces acting on a system are often so many orders of magnitude smaller than the internal forces involved (e.g., in particle interactions) that the external forces may simply be ignored. In cases where the external forces may not be ignored, or where there are substantial inflows or outflows of matter or energy, due allowance must be made for these influences when drawing up the conservation balance sheets. When all such allowances have been made, the conservation laws apply to all events and processes occurring in the universe.

Thus, the conservation laws have a special kind of universality. They are universal in the sense that the antecedent or reference class is a broad ontological category. They do not apply just to particular kinds of events or processes, but (when compensation is made for external influences) to *all* events and processes. The conservation laws thus have a scope which is wider than that of most other laws of nature. If we think of "All As are Bs" as the assertion that everything whatever in the world is either "not an A" or is "both an A and a B," then there is a sense in which this claim is a universal one. But it is not universal *in scope* in the kind of way the conservation laws are. For if something is not an event or process, it could hardly be an energy-conserving, or momentum-conserving, or lepton-number-conserving event or process. Hence the conservation laws are not like ordinary laws or generalizations. The form of a conservation law is "Every event or process which can occur is X-conservative," or equivalently, "Events and processes which are not X-conservative are impossible."

If this is the general form of a conservation law, then it is evident that it presents a serious difficulty for the DTA account of laws. The claim that events and processes which are not X-conservative are impossible is not naturally construed as an assertion of a relation between universals. For what are the universals which must be said to be related? Are energies at times, momenta at times, etc., to be counted as primitive universals? It is, surely, an advantage of our theory that it does not require the introduction of any such dubious relational properties to stand in the required necessitation relation.

It is not essential to the category of events that they should be energy-conservative, or angular-momentum-conservative, or conservative in any other respect. Changes which were not in accordance with these conservation principles would still be events. But it seems to be of the nature of the universe we live in that such events should never, or at least very rarely, occur. A universe in which the forbidden changes did occur sufficiently often would not be our kind of universe. If, for example, lepton and baryon number were not at least very nearly conserved, then electrons and protons (if they could exist at all) would not be sufficiently stable particles for matter, as we know it, to exist. Therefore, if the reason for the necessity of the conservation laws is like the reason for the necessity of other less general laws, we must suppose it to be essential to the nature of the universe that it should be one in which these quantities are at least very nearly conserved. Without the assumption that the universe is one of a kind in which certain basic quantities are conserved in this way, we do not see any satisfactory way of giving an ontological explanation of their necessity. There is no equally plausible rival bearer of essential properties which could explain the necessity of the conservation laws. Therefore, we conclude, conservation laws are best understood as ascribing properties to the world *as a whole,* properties which are essential to the natural kind to which our world belongs.

We do not claim to know for sure what the conservation laws are. But we do say that when we get them right, they will describe essential properties of the kind of world we live in. Sometimes, of course, we posit conservation laws which fail to fit the facts in the actual world. Such misdescriptions of the world are not called "laws of nature"; but they are "law-like." They are the kinds of things which would be laws, if only they were true. They describe laws which could have held, but which do not actually hold. They describe a kind of world which, if it existed, would be distinct from the one which actually exists. There are many kinds of worlds which it is logically possible for there to have been; and that is why conservation laws are contingent. But what we are seeking, when we make conjectures about the conservation laws, are truths about the essential properties of this kind of world. It is to be hoped that the logically possible kinds of worlds we are describing will be getting closer and closer to the kind to which this world does in fact belong.

Conservation laws are especially instructive, because they lend themselves exceedingly eagerly to our general analysis of laws. Conservation laws do look, on the face of things, like descriptions of essential properties of the world as a whole. It takes an effort to rewrite them in such a way that they sound as though they are describing correlations of some parts of the world with others. It takes somewhat less effort, but some effort nevertheless, to rewrite conservation laws in such a way that they sound as though they are describing essential properties of mere proper parts of the world. Yet if you take conservation laws to assign essential properties to the kind of world we live in, then they can be taken more or less at face value. They fall into place without rewriting. Conservation laws are not, however, the only laws which fall into place neatly, if construed as describing the natural kind to which the whole world belongs. We urge, in fact, that *all* laws of nature are best understood in this way.

On present evidence, it seems that the kind of world we live in is a Minkowskian world with a four-dimensional, space-time structure. The laws of relativity and gravitation are of the essence of this structure. Worlds like ours are ones which display certain global symmetries. The symmetry principles derive from this. The world is one of a kind which consists basically of a relatively small number of kinds of fundamental particles or fields. The fundamental particles are known to occur in families, and the different kinds of fields are thought to be related. It is implausible that their nature should turn out to be independent of each other. We speculate, therefore, that when the fundamental nature of the world is better understood, it will be predictable what kinds of fundamental particles and fields there are, and what their essences must be. The particles or fields which exist in the world change and interact with each other in all of the ways permitted by the conservation laws. The interactions that can occur must do so with probabilities determined by the appropriate solutions to the field equations for the kinds of fields involved. This is the source, we suppose, of the probabilistic laws of quantum mechanics. This is the kind of way we think the various laws of nature are to be understood as arising out of essences. They all derive ultimately from the nature of the world we inhabit.

Our theory of laws gives laws a kind of necessity which is quite distinct from logical necessity. It permits laws to be logically contingent, and yet also necessarily true—necessarily true of any world of the same natural kind as our world. It also grants laws of nature the right kind of universality. Essences of things in the world, and correlations which depend on the essences of such things may both contribute to the essence of the world as a whole. For the world may be such that it necessarily contains things of these kinds, which consequently must be correlated in these ways. That is why some truths about essences of parts of the world, such as the laws of electromagnetism, may properly be regarded as laws of nature. They contribute to the essence of the world as a whole. Essential properties of parts of the world, and correlations among parts of the world, constitute laws of nature just when they give us general information about the essential properties of the kind of world we live in.

Notes

1. Our theory is not entirely unanticipated by the work of others. Harré and Madden [1975] have advanced a perspective on laws of nature akin to ours.

2. Dretske [1977]; Tooley [1977], [1987]; Armstrong [1978], [1983].

3. They are from the same genre insofar as all three accounts describe laws in terms of necessary connections between universals, but they differ from one another in the way each theory is formulated. For instance, one important difference between Tooley's and Armstrong's accounts concerns the ontology of universals. Armstrong, unlike Tooley, does not allow uninstantiated universals. While such differences may be significant when comparing the three theories with each other, these nuances need not concern us here.

4. Critical discussions of Armstrong's, Dretske's, and Tooley's accounts can be found in Carroll [1987]; Earman [1984]; Forge [1986]; Hetherington [1983]; Lewis [1983]; Niiniluoto [1978]; and van Fraassen [1987]. For some replies see Armstrong [1988]; Dretske [1978]; and Tooley [1987].

5. Shoemaker [1980] and Swoyer [1982] defend similar versions of a "bottom-up" approach.

6. Lewis [1983], p. 366.

7. John Carroll raises this possible approach and following objection in his [1987], p. 265.

8. As defined by Peter Forrest [1988], p. 5.

9. This objection was brought to our attention by Freya Mathews, John Bacon, and John Fox.

10. Armstrong is careful to make a distinction between the task of discovering what sorts of things or properties there are in the universe and how they are constituted, and the separate exercise of finding what laws link these properties together. In this framework, the discovery that copper has the atomic number 29 is classed as belonging to the first enterprise, as it is a truth about the internal structure or "geography" of a universal. While Armstrong states that each enquiry is inextricably bound up with each other, he nevertheless argues that they are distinguishable (see Armstrong [1983], pp. 3 and 139).

11. The magnetic field is in fact always non-divergent; the electric field is non-divergent in every region except where there are electric charges.

12. The electrostatic and magnetostatic fields are the special cases where the magnetic and the electric field gradients respectively are zero.

References

Armstrong, D. M. [1978]: *Universals and Scientific Realism,* 2 vols. Cambridge: Cambridge University Press.

Armstrong, D. M. [1983]: *What Is a Law Nature?* Cambridge: Cambridge University Press.

Armstrong, D. M. [1988]: "Discussion: Reply to van Fraassen," *Australasian Journal of Philosophy* 66, pp. 224–29.

Carroll, John W. [1987]: "Ontology and the Laws of Nature," *Australasian Journal of Philosophy* 65, pp. 261–76.

Dretske, F. I. [1977]: "Laws of Nature," *Philosophy of Science* 44, pp. 248–68.

Dretske, F. I. [1978]: "Discussion: Reply to Niiniluoto," *Philosophy of Science* 45, pp. 440–44.

Earman, J. [1984]: "Laws of Nature: The Empiricist Challenge," in R. J. Bogdan (ed.), *D. M. Armstrong.* Dordrecht: Reidel.

Forge, J. [1986]: "Discussion: David Armstrong on Functional Laws," *Philosophy of Science,* 53, pp. 584–87.

Forrest, P. [1988]: "Supervenience: The Grand-Property Hypothesis," *Australasian Journal of Philosophy* 66, pp. 1–12.

Fraassen, B. Van [1987]: "Armstrong on Laws and Probabilities," *Australasian Journal of Philosophy* 65, pp. 243–60.

Harré, R. and Madden, E. H. [1975]: *Causal Powers: A Theory of Natural Necessity.* Oxford: Basil Blackwell.

Hetherington, S. [1983]: "Tooley's Theory of Laws of Nature," *Canadian Journal of Philosophy* 13, pp. 101–6.

Lewis, D. [1983]: "New Work for a Theory of Universals," *Australasian Journal of Philosophy* 61, pp. 433–39.

Niiniluoto, I. [1978]: "Discussion: Dretske on Laws of Nature," *Philosophy of Science* 45, pp. 431–39.

Shoemaker, S. [1980]: "Properties and Causality," in P. van Inwagen (ed.), *Time and Cause.* Dordrecht: Reidel.

Swoyer, C. [1982]: "The Nature of Natural Laws," *Australasian Journal of Philosophy* 60, pp. 203–23.

Tooley, M. [1977]: "The Nature of Laws," *Canadian Journal of Philosophy* 7, pp. 667–98.

Tooley, M. [1987]: *Causation: A Realist Approach.* Oxford: Clarendon Press.

9

Natural Laws and the Problem of Provisos

MARC LANGE

I

According to the regularity account of physical law—versions of which have been advocated by Ayer (1963), Braithwaite (1953, ch. 9), Goodman (1983, pp. 17–27), Hempel (1965a, pp. 264ff.), Lewis (1973, pp. 72–77; 1986), Mackie (1962, pp. 71–73), Nagel (1961, pp. 58ff.), and Reichenbach (1947, ch. 8), among others—laws of nature are regularities among events or states of affairs and a law-statement, the linguistic expression of a law, is a description of a regularity that is a law. The familiar challenge faced by this account is to distinguish those descriptions of regularities that are law-statements from those that are accidental generalizations. I wish to consider a more fundamental problem: that many a claim we believe to describe no regularity at all, nomological or accidental, we nevertheless accept as a law-statement. This problem arises from what Hempel (1988) calls "the problem of provisos."

Consider the familiar statement of the law of thermal expansion: "Whenever the temperature of a metal bar of length L_0 changes by ΔT, the bar's length changes by $\Delta L = k \cdot L_0 \cdot \Delta T$, where k is a constant characteristic of that metal." This statement states a relation between L_0, ΔT, and ΔL that does not obtain; it may be violated, for instance, if someone is hammering the rod inward at one end. Since this statement does not describe a regularity, it is not a law-statement, on the regularity account.

Hempel (1988) would hold that a complete statement of the law includes the condition ". . . if the end of the rod is not being hammered in." On this view, whereas the familiar "law-statement" takes the form $(x)(Fx \supset Gx)$, and the only premise one must add to yield the conclusion Ga is Fa, the genuine law-statement takes the form $(x)(Fx \: \& \: Px \supset Gx)$, and one

From *Erkenntnis* 38 (1993): 233–48.

needs to add *Pa* as well as *Fa* to infer *Ga.* It is the regularity account of laws that leads Hempel to characterize such conditions (*x*) . . . if *Px* as "essential" to law-statements (and the corresponding premises *Pa* as "essential" to inferences); without these conditions, the claims would be false and so, on the regularity account, would not be law-statements. Following Hempel (1988, p. 23), I'll refer to those conditions (and premises) that, by this reasoning, are necessary to law-statements (and to inferences from law-statements), but are "generally unstated," as "provisos."

Provisos pervade scientific practice. By Hempel's reasoning, Snell's law of refraction—when a beam of light passes from one medium to another, $\sin i/\sin r = \text{constant}$, where *i* is the angle of incidence of the beam upon the second medium, *r* is the angle of refraction in that medium, and the constant is characteristic of the two types of media—must require particular temperatures and pressures of the media as well as the absence of any magnetic or electrical potential difference across the boundary, uniform optical density and transparency and non-double-refractivity in the two media, and a monochromatic beam; these conditions are provisos. Likewise, the law of freely falling bodies—the distance a body falls to earth in time *t* is $(1/2)gt^2$—must specify when fall qualifies as "free"; while the law can remain approximately true away from the height at which *g* is measured, its predictions may be drastically wrong when electromagnetic forces, air resistance, or other collisions affect the falling body.

On Hempel's proposal, it becomes impossible to state very many genuine law-statements since, as Giere puts it (1988, p. 40), "the number of provisos implicit in any law is indefinitely large." To state the law of thermal expansion, for instance, one would need to specify not only that no one is hammering the bar inward at one end, but also that the bar is not encased on four of its six sides in a rigid material that will not yield as the bar is heated, and so on. For that matter, not all cases in which the bar is hammered upon constitute exceptions to $\Delta L = k \cdot L_0 \cdot \Delta T$; the bar may be hammered upon so softly and be on such a frictionless surface that the hammering produces translation rather than compression of the bar. One is driven to say that the only way to utter a complete law-statement is to employ some such condition as ". . . in the absence of other relevant factors." But Hempel deems such an expression inadmissible in a law-statement: On the regularity account, a law-statement states a particular relation (which must obtain and be a law), but a claim (*x*) (*Fx* & there are no other relevant factors ⊃ *Gx*) does not assert any determinate relation at all because it fails to specify which other factors count as relevant, i.e., which specific premises are to be added to *Fa* and the law-statement to infer *Ga*. Such a claim is no better than "The relation $\Delta L = k \cdot L_0 \cdot \Delta T$ holds when it holds," and so is further from being a law-statement than is the familiar statement of the law of thermal expansion, which at least ascribes a particular (albeit false) relation to various quantities.

In short, Hempel sees the existence of provisos as posing a dilemma: For many a claim that we commonly accept as a law-statement, either that claim states a relation that does not obtain, and so is false, or is shorthand for some claim that states no relation at all, and so is empty. In either case, the regularity account must admit that many claims commonly accepted as law-statements are neither complete law-statements themselves nor even colloquial stand-ins for complete law-statements. If we continue to regard those familiar claims as law-statements, then we violate the regularity account. This is the problem of provisos. (Prior to Hempel 1988, versions of this problem were discussed by Canfield and Lehrer 1961 and Coffa 1968, as well as by Hempel himself 1965b, p. 167, and the issue, in general terms, was anticipated by Scriven 1961. Difficulties similar to the problem of provisos have also been noted in ethics, e.g., with regard to Ross's 1930 definition of a "prima facie duty.")

One may be tempted to reject this problem by insisting that genuine law-statements lack provisos; on this view, that many familiar "law-statements" are actually neither law-statements nor abbreviations for law-statements only goes to show that we have discovered very few genuine laws (or nomological explanations). To yield to this temptation would, I think, be unjustified. An account of laws must accommodate the fact that scientists show no reluctance to use these familiar claims in the manner distinctive of law-statements, e.g., in explanations and in support of counterfactuals. This fact would be difficult to explain if we held that scientists do not consider them to express laws. To insist nevertheless that only claims without provisos are genuine law-statements is to hold that whether a claim must or must not be captured by an analysis of what it is to believe a claim to state a physical law is not determined by whether scientists treat that claim as able or unable to perform those functions that distinguish law-statements from accidental generalizations. But, then, on what basis is the adequacy of an account of law to be evaluated? Scientific practice is the only phenomenon that exists for an account of law to save; if an account is tested against not actual science but science as idealized to conform to that account, the test is circular. If accounts of law are not free to disregard the fact that scientists treat "All gold cubes are smaller than one cubic mile" as an accidental generalization and "All cubes of Uranium-235 are smaller than one cubic mile" as expressing a law, even though this fact is troublesome for many proposed accounts, I see no reason why an account of law should be permitted to ignore the fact that scientists treat as law-statements many claims involving provisos.

Another temptation is to argue that although some law-statements involve provisos, these derive from other, more fundamental law-statements, which themselves need no provisos to describe regularities. For instance, the law of falling bodies follows in classical physics from the fundamental laws of motion and gravitation along with information about the earth; the proviso ". . . so long as the body is falling freely" restricts the law to those cases in which the gravitational-force law applied to this information about the earth accounts for all of the forces acting on the body. However, this temptation should also be resisted. It merely pushes the burden onto the fundamental laws: What regularity is described by the gravitational-force law? If it described a regularity between the masses and separation of two bodies and the *total* force that each exerts on the other, it would be false unless it included the proviso ". . . so long as the bodies exert no other forces upon each other." But to regard this proviso as part of this "fundamental" law not only conflicts with the law's applicability to a case in which two charged and massive bodies interact, but also raises the familiar problem: The proviso fails to specify the circumstances in which other forces are present, and so prevents the gravitational-force law from setting forth a particular relation. If each of the other force-laws is supposed to specify when a given non-gravitational force is present and thereby help to determine the relation asserted by the gravitational-force law, then each of these other force-laws would have to apply to a body affected by many types of forces and so could not describe a regularity involving the *total* force exerted on the body. Alternatively, for the gravitational-force law to describe a regularity between the masses and separation of two bodies and the *gravitational* (rather than total) force that each exerts on the other, a component gravitational force would have to be a real entity that conforms to certain regularities. That component forces are real is a controversial contention (see, e.g., Cartwright 1983), and in any event, it seems to me that the nomic status of the gravitational-force law does not depend on it; after all, we use the Coriolis-force law in the manner distinctive of law-statements even though we believe there to be no Coriolis force to figure in a regularity that the law-statement might describe.

Perhaps one would be justified in setting the problem of provisos aside if one had some account of physical law that, except for this problem, were entirely successful. But since I know of no such account, I think it worth investigating whether greater progress toward one can be made by reflecting on provisos than by disregarding them. In this paper, I'll offer a response to the problem of provisos that ultimately undermines the regularity account of physical law and suggests an alternative, normative conception of law-statements according to which they specify the claims we ought to use, in various contexts, to justify certain other claims. I'll argue that some claims are properly adopted as law-statements although they are believed not to describe regularities, because one who believes that "It is *F*" ought to be used to justify "It is *G*" need not believe that some regularity, such as that all *F*s are *G*, obtains.

In Section II, I'll explain the problem of provisos more fully, and in Section III, I'll attack Hempel's view of what a "law-statement" must be to qualify as complete, on which the problem depends. The more liberal criterion of completeness that I'll defend permits us to avoid Hempel's dilemma by enabling us to regard familiar "law-statements" as law-statements. I'll argue in Section IV, however, that this response to the problem of provisos requires the rejection of any regularity account of physical law because many a claim commonly accepted as a law-statement describes no regularity; a normative conception of law-statements then suggests itself. In Section V, I'll maintain that this strategy can be used to argue against many other conceptions of physical law besides the regularity account, such as those of Armstrong (1983) and Kneale (1949). Finally, I'll argue that my response to the problem of provisos is superior to that offered by Giere (1988).

II

Hempel does not explicitly present the problem of provisos as posing the dilemma that "law-statements" are either false or empty. Yet this dilemma certainly stands behind his discussion. It must be because he believes a "law-statement" without provisos would be false, and so would not be a law-statement, that Hempel defines provisos as "essential." He goes on to point out that if the complete law-statement includes the proviso (*x*) . . . if

Px, then among the premises of an inference from the law-statement to testable predictions must be *Pa,* as well as other auxiliary hypotheses. This might at first appear to reduce at least part of the problem of provisos to a special case of the Duhem-Quine problem: a law-statement is not falsifiable (at least, not in a straightforward sense) because to make a prediction from it that can be tested, one must use auxiliary hypotheses, so one can preserve the law-statement, if the prediction fails, by rejecting an auxiliary hypothesis. If provisos are merely additional auxiliary hypotheses, distinguished from others only by the fact that they generally go unstated in scientific practice, then (it might appear) they do not represent a novel kind of threat to the falsifiability of individual hypotheses.

Hempel (1988, pp. 25f.) emphasizes, however, that the existence of provisos presents some obstacle to falsifiability beyond that posed by the Duhem-Quine problem. The Duhem-Quine problem assumes that the law-statements and auxiliary hypotheses are jointly sufficient to entail the testable prediction. But, Hempel maintains, if there are provisos among the auxiliary hypotheses, then this assumption often fails because, for many a familiar "law-statement," one can state neither the complete set of proviso conditions (*x*) . . . if *Px* needed to make that "law-statement" true nor the complete set of auxiliary hypotheses *Pa* needed to infer the testable prediction from the law-statement. Therefore, Hempel says that in comparison to the Duhem-Quine problem, "[t]he argument from provisos leads rather to the stronger conclusion that even a comprehensive system of hypotheses or theoretical principles will not entail any [testable predictions] because the requisite deduction is subject to provisos"—that is to say, always remains subject to provisos, no matter how many auxiliary premises one adds to try to exclude all factors disturbing to the law.

Hence, it is only because he considers complete law-statements impossible, since "the number of provisos implicit in any law is indefinitely large" (Giere 1988, p. 40), that Hempel sees provisos as presenting a difficulty that is distinct from the Duhem-Quine problem. But suppose Hempel believed that a condition such as ". . . in the absence of other relevant factors" could appear in a law-statement, and likewise that "There are no other relevant factors" could function as an auxiliary hypothesis. Then he would have to admit that any law *can* be completely stated by a claim that includes only a finite number of conditions (*x*) . . . if *Px,* one of which might be ". . . if there are no disturbing factors," and that any inference to a testable prediction includes only a finite number of auxiliary hypotheses *Pa,* one of which might be "There are no disturbing factors." The existence of provisos would then add nothing new to the Duhem-Quine problem. Hempel therefore must regard the expression ". . . in the absence of other relevant factors" as inappropriate for a law-statement. Though he does not explain why this is, the regularity account of laws, implicit throughout his discussion, suggests an answer: The sentence "$\Delta L = k \cdot L_0 \cdot \Delta T$ obtains in the absence of factors that disturb it" states no definite relation and so cannot be a law-statement. This worry is evident in Giere's remark (1988, p. 40): "The problem is to formulate the needed restrictions without rendering the law completely trivial."

III

By Hempel's definition, a proviso (x) . . . if Px usually is omitted from a statement of the law, and the corresponding premise Pa usually is not mentioned in inferences involving that law. Hempel regards these inferences as enthymemes and these familiar "law-statements" as incomplete. But why are we able to make do with incomplete law-statements? And is Hempel correct in considering them incomplete? I'll now argue that Hempel's criterion of completeness is motivated by an incorrect view of what is necessary in order for a sentence to state a determinate relation. I'll argue that familiar law-statements, which include clauses such as "in the absence of other relevant factors," are complete as they stand.

That proviso premises are distinguished by their absence from ordinary conversation is not an incidental feature of Hempel's definition. As I've explained, it is bound up with the fact that the number of provisos is "indefinitely large," which makes it impossible to offer them *all* as premises. But why do scientists find it unnecessary to mention *any* of the proviso premises in order to put an end to demands for the justification of their conclusions? Why is it that although it is known that when someone is hammering on the bar, $\Delta L = k \cdot L_0 \cdot \Delta T$ need not obtain, in actual practice a claim concerning k, L_0, and ΔT is recognized as sufficient, without "No one is hammering on the bar," to put an end to demands for the justification of a claim concerning ΔL?

The answer is that in practice, when no one is hammering on the bar, nearly all of those who demand justifications of claims concerning ΔL already believe that this is so. (Likewise, to consider a different proviso example, it is widely understood by workers in many fields of physics that no cases will involve velocities approaching that of light.) Of course, to someone who presents a claim concerning *k*, L_0, and ΔT to justify a claim concerning ΔL, one *could* object, "You have not told me that no one is hammering on the bar, and this you must do because if someone is, then (you will agree) your conclusion may well be false even though your premise is true." But apparently, that no one is hammering on the bar would, in nearly any actual case in which it is true, be believed in advance by those who might demand the justification of some claim concerning ΔL. In nearly all cases, then, someone who demands a justification for a claim concerning ΔL should regard a claim concerning k, L_0 and ΔT as a sufficient response.

Attention to this kind of shared background not only explains why scientists needn't in practice give any of the "indefinitely large" number of proviso premises in order to justify their conclusions, but also reveals why Hempel's standard of completeness is too high. Hempel apparently considers a "law-statement" complete only if it suffices, in the absence of any background understanding, to inform one of what it takes for nature to obey the corresponding law. This ideal of completeness requires that the complete law-statement include all of the proviso conditions (*x*) . . . if *Px* and, more importantly, that none of these conditions be "in the absence of disturbing factors," because this condition plainly appeals to background understanding. A generalization "All *F*s are *G*, except when disturbing factors are present" does not indicate, in a manner intelligible to one who doesn't know already what constitutes a disturbing factor, some determinate regularity to which nature conforms. In exactly the same way, someone who is told to follow a rule "Conform to the regularity . . . when it is appropriate to apply this rule" can understand what it would take to follow this rule only if she already knows when this rule is appropriately applied.

But to require that a rule be intelligible in the absence of implicit background understanding of how to apply it is not a reasonable criterion of completeness because no rule can satisfy it. As Wittgenstein (1958) suggests, one can always conceive of alternative interpretations of a rule that recognize the same actual past actions as conforming to the rule but regard different hypothetical actions as what it would take to follow the rule. It is futile to try to avoid this by including in the rule an expression that specifies explicitly how to apply the rest of the rule, for alternative interpretations of that expression are likewise conceivable. In the same way, a law-statement specifies a determinate relation only by exploiting implicit background understanding of what it would take for nature to obey this law.

This point applies to *any* law-statement, whether or not it *blatantly* appeals to implicit background understanding by referring to "disturbing factors."

That the proper way to apply a rule is not itself specified by any rule, intelligible without implicit background understanding of how to follow that rule, does not imply, as Hempel seems to think, that nothing counts as a violation of the given rule. The background understanding, albeit implicit, enables the rule to impose determinate requirements. This implicit understanding must be capable of being taught and of being made the explicit subject of discussion if disputes over it ever arise. Some fortune tellers explain away your failure to make accurate predictions by using their rules as the result of your having misapplied the claims they believe to be law-statements, of your having ignored some clause they say they neglected to mention. It is doubtful that they have undertaken any determinate commitments at all by adopting those "law-statements." It is as if someone says, "I can run a four-minute mile," but with each failure reveals a proviso that she had not stated earlier: ". . . except on this track," ". . . except on sunny Tuesdays in March," and so on. It quickly becomes apparent that this person will not acknowledge having committed herself to any claim by asserting "I can run a four-minute mile." Science is distinguished from such bunk neither by the explicit inclusion in scientific law-statements of all conditions Hempel would deem "essential" nor by the absence of implicit background understanding of how to apply those law-statements. What is noteworthy about science is that this background understanding is genuine background *understanding*. In general, all researchers identify the same testable claims as those to which one would become committed by adding a given lawlike hypothesis to a certain store of background beliefs. Because they agree on how to apply the hypothesis, it is subject to honest test.

IV

I'll now argue that this attractive response to the problem of provisos is incompatible with the view that law-statements describe regularities of a certain kind. This account of provisos leads instead to a conception of law-statements as specifying the claims we ought to respect, in a certain context, as able to justify certain other claims.

What, according to the regularity account, is the regularity stated by the familiar expression of the law of thermal expansion? Presumably, it is that a bar's length changes by $k \cdot L_0 \cdot \Delta T$ whenever the bar has a certain composition, its initial length is L_0, its temperature changes by ΔT, and there are no disturbing factors. That nearly all of us agree on whether "There are no disturbing factors" is appropriately said of a given case saves this expression, and so the law-statement, from emptiness; we share an implicit understanding of which predictions the law-statement underwrites, of when it is properly applied. But whether certain scientists are correct in saying of a given case that there are no disturbing factors depends not just on the physical features of this case but also on their purposes, e.g., on the degree of approximation they can tolerate considering the use they intend to make of this prediction. Even if the regularity account can countenance as laws some uniformities involving the concerns of scientists, the law of thermal expansion was surely not supposed to be such a uniformity; somehow, the subject has changed from a law of physics to a law of the science of scientific activity.

It gets worse for the regularity account. Suppose that according to the regularity account, the familiar expression of the law of thermal expansion states that a bar's length changes by $k \cdot L_0 \cdot \Delta T$ whenever the bar has a certain composition, its initial length is L_0, its temperature changes by ΔT, no one is hammering on the bar hard enough to cause deviations from $\Delta L = k + L_0 + \Delta T$ great enough (given our interests) to matter to us, and so on. Nevertheless, this claim cannot qualify as a law-statement according to the regularity account, for the relation it states does not obtain. It is violated, for example, when someone is hammering on the bar hard enough to cause the actual change in the bar's length to depart from $k \cdot L_0 \cdot \Delta T$ but lightly enough for this departure to be irrelevant to the investigator's concerns. While in this case it is not true that the actual length of the bar changes by $k \cdot L_0 \cdot \Delta T$, it is proper for such an investigator to predict that the length of the bar will change by $k \cdot L_0 \cdot \Delta T$. The relation stated by the law involves not the bar's actual change in length but rather the change in length one is justified in predicting.

This suggests that the law of thermal expansion doesn't describe a regularity among events or states of affairs but concerns the way a claim concerning ΔL ought to be justified. The response I've advocated to the problem of provisos leads to a conception of the law of thermal expansion as the objective fact that under certain (partly pragmatic) circumstances, a premise about the bar's initial length, its change in temperature, and its composition ought to be used to justify a certain claim about its change in length. On this view, a law-statement has a normative element because it says that under certain circumstances, certain claims ought to be used to justify certain other claims.

To succeed, this account would have to show that law-statements are able to explain their instances, to support counterfactuals, and to be confirmed inductively by their instances *because* they specify the roles that certain descriptions should play in justifications. This account would likewise have to show that *because* accidental generalizations lack this prescriptive import, they cannot be used in these ways. To show this would be to break the familiar unilluminating circle of analysis from a law's explanatory power, to its physical necessity, to its capacity for counterfactual support, to its lawlikeness, to its capacity to be inductively confirmed by its instances, to its explanatory power. This task is well beyond the scope of this paper; I begin it in my (1993) article. My concern here is to show how this normative account of law-statements arises from a plausible response to the problem of provisos.

Let me summarize the argument. Contrary to Hempel, a law-statement that includes "so long as there are no disturbing factors" is not thereby rendered trivial. Like any other expression, this condition derives its content from an implicit shared understanding of how one should use it. Hence, the relation that a "law-statement" that includes this condition claims to obtain, which is determined by the appropriate way to apply this statement, ultimately depends on proprieties not codified explicitly. All there is to give meaning to "so long as there are no disturbing factors," and thereby save the "law-statement" from triviality, is how scientists consider the "law-statement" properly applied; because there is near unanimity on this point, the "law-statement" is not empty. But since the statement is properly applied to a given physical circumstance only when investigators have certain interests, the regularity that supposedly constitutes the law described by this statement must involve not only physical events but also investigators' concerns. Even if the regularity account regarded such uniformities as laws of physics, these uniformities do not obtain anyway and so, according to the regularity account, cannot be described by law-statements. For with regard to a given physical situation, it sometimes is and sometimes is not appropriate for us to say that ΔL will be $k \cdot L_0 \cdot \Delta T$, depending on our concerns. But surely in a given physical situation, there is only one real amount by which the bar expands; the actual behavior of the bar does not depend on our interests. Once the law-statement, in stating a relation involving "no one is hammering on the bar hard enough or otherwise disturbing it enough to matter to us," turns out to involve our interests, the other relatum, "ΔL," is found to be infected by our interests as well. It refers to the change in the bar's length according to the claim that ought to be considered justified by these premises, which depends on the purpose for which we intend to use this claim, rather than to the bar's actual change in length, which does not. So the law-statement expresses a norm rather than a regularity involving the bar's real length. It states a relation between the presence of certain conditions (some having to do with our interests) and the way one ought to predict the bar's change in length.

A law-statement, then, informs an audience already able to tell whether there are "disturbing factors" that if there are none, they ought to use a given claim to justify another. On this view, a proviso is not "essential" *tout court,* contrary to Hempel. There is no fact of the matter to whether a law-statement has or lacks provisos. There is only whether those who would typically discuss that law-statement can learn, merely from reading it, what it prescribes they do (as is the case for the familiar expression of Newton's second law), or whether making the proviso explicit is essential *for that audience.*

Moreover, a normative conception of law-statements does not deny that there are regularities in nature. It denies only that law-statements, when performing their distinctive functions, are describing some of them. Laws, in the sense that the regularity account envisions them, need not be invoked to understand what law-statements say.

V

The foregoing argument, if successful, can be used to undermine not only any regularity account of law but also any account that takes a law-statement to describe a state of affairs that necessarily presupposes such a regularity. For instance, it has been suggested (e.g., by Kneale [1949]) that law-statements describe not regularities of a certain kind but non-Humean connections of physical necessity. Such an account implies that the non-Humean connection supplements a regularity among circumstances. But with this regularity, the problem of provisos takes hold. The same reasoning can be deployed against recent accounts (see Armstrong 1983, and Dretske 1977) according to which law-statements describe relations among universals, such as the property of lengthening by a given amount. On these accounts, the law-statement of thermal expansion entails that events conform to a certain relation "so long as there are no disturbing factors." Precisely what this comes to, i.e., whether some factor qualifies as "disturbing" or not, must be fixed by the law-statement. But, I have argued, the law-statement appeals to a determinate set of disturbing conditions only because it states a relation involving not the property of expanding in length by a given amount but the property of being able to serve as the subject of a justified claim attributing expansion by a given amount. In short, if the above reasoning goes through, it constitutes a recipe for an argument against any account according to which a law is or requires a regularity among events or states of affairs.

A law-statement concerning a particular influence, such as the Coriolis-force law or Newton's two-body gravitational-force law, suffices to tell persons *having comparatively little background understanding* how they ought to justify a certain claim. This is because the magnitude of the influence covered by the law-statement does not depend on which other influences are at work; hence, while there are provisos, there are none that demand significant background understanding, such as "in the absence of disturbing factors" would. The gravitational-force law, for example, specifies how one should justify a claim concerning the gravitational force between two bodies, whatever the other influences with which this subtotal should be combined to reach, say, the total force on a given body.

However, Giere's (1988) response to the problem of provisos does not capture the fact that to be committed to the gravitational-force law is to be committed to treating a certain inference to the component gravitational force as correct, no matter what the other relevant influences. In light of the problem of provisos, Giere holds certain familiar law-statements to be false as claims about the physical world, but he contends that, as law-statements, they must nevertheless function as descriptions of something. He therefore tries to find something that they describe. While this search has led Armstrong and Dretske to the exotic realm of universals, Giere (along with Cartwright 1983, whose account is similar in all relevant respects) maintains that a law-statement describes a scientific model. One may hold that the relevant behavior of a given real system can be predicted by using some model; Giere terms such a claim a "theoretical hypothesis." Since the law-statement describes the model, not reality, it needs no qualification by provisos to be accurate.

Consider, then, what Giere says (1988, p. 44) about two laws, each concerning a single influence, that are combined to account for a magnetically influenced pendulum:

> We have discovered a new kind of pendulum . . . in which the force of gravity is supplemented by a magnetic force directed toward a point below the point of rest. Constructing a theoretical model that does apply to such systems is a fairly easy problem in physics.

If the gravitational-force law specifies how one ought to calculate a subtotal (the gravitational influence of one body on another) no matter what the other influences present, and the magnetic-force law does likewise, then this is indeed an easy problem. Having already accepted these law-statements, and having recognized the proper way to add forces and to use the net force on a body to infer its acceleration, we are committed to a particular procedure for predicting the bob's motion.

But Giere takes these law-statements not to prescribe which models to use (which is the job of theoretical hypotheses) but merely to describe certain models. Thus, Giere must admit that by accepting that the gravitational-force law describes certain models, we are not committed to saying that one ought to use a model it describes to predict the bob's motion. Moreover, one who accepts the theoretical hypothesis that this law should be used to calculate the gravitational force exerted by the earth on a non-magnetic bob is not thereby committed to the theoretical hypothesis that this law should be used to calculate the gravitational force exerted by the earth on a magnetically influenced bob.

On my view, in contrast, to accept this law-statement is to recognize the way one should justify claims concerning the component gravitational force, whatever the other relevant influences. This view accounts for what Giere's view obscures: That we are committed to some common element in our treatments of ordinary and magnetically augmented pendula (namely, an identical way of justifying a certain subtotal) in virtue of which the magnetically augmented pendulum is an easy problem. We became committed to elements of its solution when we adopted solutions to other problems.

I have argued against Hempel's contention that a complete law-statement must specify its own range without depending on implicit background understanding in order to do so. I have also argued against Giere's (and Cartwright's) alternative claim that a law-statement says nothing about its range, leaving it for theoretical hypotheses to specify. I have defended the view that a law-statement specifies its range in a fashion that may involve blatant appeal to implicit proprieties of use. I have thereby tried to offer a way around the problem of provisos, at the price of abandoning the regularity account of law in favor of a normative analysis. Further discussion of that proposal must await another occasion.

References

Armstrong, D.: 1983, *What Is a Law of Nature?* Cambridge University Press, Cambridge.

Ayer, A. J.: 1963, "What Is a Law of Nature?" in *The Concept of a Person and Other Essays,* Macmillan, London, pp. 209–34.

Braithwaite, R. B.: 1953, *Scientific Explanation,* Cambridge University Press, Cambridge.

Canfield, J., and Lehrer, K.: 1961, "A Note on Prediction and Deduction," *Philosophy of Science* 28, 204–8.

Cartwright, N.: 1983, *How the Laws of Physics Lie,* Clarendon Press, Oxford.

Coffa, J.: 1968, "Discussion: Deductive Predictions," *Philosophy of Science* 35, 279–83.

Dretske, F.: 1977, "Laws of Nature," *Philosophy of Science* 44, 248–68.

Giere, R.: 1988, "Laws, Theories, and Generalizations," in A. Grunbaum and W. Salmon (eds.), *The Limits of Deductivism,* University of California Press, Berkeley, pp. 37–46.

Goodman, N.: 1983, *Fact, Fiction, and Forecast,* 4th ed., Harvard University Press, Cambridge.

Hempel, C. G.: 1965a, "Studies in the Logic of Explanation," in *Aspects of Scientific Explanation,* The Free Press, New York, pp. 245–95.

Hempel, C. G.: 1965b, "Typological Methods in the Natural and the Social Sciences," in Ibid., pp. 155–72.

Hempel, C. G.: 1988, "Provisos," in A. Grunbaum and W. Salmon (eds.), *The Limits of Deductivism,* University of California Press, Berkeley, pp. 19–36.

Kneale, W.: 1949, *Probability and Induction,* Oxford University Press, Oxford.

Lange, M. B.: 1993, "Lawlikeness," *Noûs* 27: 1–21.

Lewis, D. K.: 1973, *Counterfactuals.* Harvard University Press, Cambridge.

Lewis, D. K.: 1986, "Postscript to 'A Subjectivist's Guide to Objective Chance,'" in *Philosophical Papers: Volume 2,* Oxford University Press, Oxford, pp. 121–26.

Mackie, J. L.: 1962, "Counterfactuals and Causal Laws," in R. S. Butler (ed.), *Analytic Philosophy,* Barnes and Noble, New York, pp. 66–80.

Nagel, E.: 1961, *The Structure of Science,* Harcourt, Brace, and World, New York.

Reichenbach, H.: 1947, *Elements of Symbolic Logic,* Macmillan, New York.

Ross, W. D.: 1930, *The Right and the Good,* Oxford University Press, Oxford.

Scriven, M.: 1961, "The Key Property of Physical Laws—Inaccuracy," in H. Feigl and G. Maxwell (eds.), *Current Issues in the Philosophy of Science,* Holt, Rinehart, and Winston, New York, pp. 91–104.

Wittgenstein, L.: 1958, *Philosophical Investigations,* 3rd ed., trans. G. E. M. Anscombe, Macmillan, New York.

10

Humean Supervenience

BARRY LOEWER

Over the last couple of decades David Lewis has been elaborating and defending a metaphysical doctrine he calls "Humean Supervenience" (HS). Here is how he introduces it.

> Humean supervenience is named in honor of the great denier of necessary connections. It is the doctrine that all there is to the world is a vast mosaic of local matters of particular fact, just one little thing and then another. . . . We have geometry: a system of external relations of spatiotemporal distances between points. . . . And at those points we have local qualities: perfectly natural intrinsic properties which need nothing bigger than a point at which to be instantiated. For short: we have an arrangement of qualities. And that is all. There is no difference without difference in the arrangement of qualities. All else supervenes on that.[1]

In this paper I explore and to an extent defend HS. The main philosophical challenges to HS come from philosophical views that say that nomic concepts—*laws, chance,* and *causation*—denote features of the world that fail to supervene on non-nomic features. Lewis rejects these views and has labored mightily to construct HS accounts of nomic concepts. His account of laws is fundamental to his program, since his accounts of the other nomic notions rely on it. Recently, a number of philosophers have criticized Lewis's account, and Humean accounts of laws generally, for delivering, at best, a pale imitation of the genuine item.[2] These philosophers think that the notion of law needed by science requires laws—if there are any—to be fundamental features of our world that are completely distinct from and not supervenient on the particular facts that they explain. I side with Lewis against these philosophers. Here I will argue that although Lewis-laws don't fulfill *all* our philosophical expectations, they do play the roles

From *Philosophical Topics* 24 (1996): 101–27.

that science needs laws to play. The metaphysics and epistemology of Humean laws, and more specifically, Lewis-laws, are in much better shape than the metaphysics and epistemology of the main anti-Humean alternatives. However, I do have misgivings about Lewis's account. Both he and his critics assume that the basic properties are so individuated so that the laws are not metaphysically necessary. If this assumption is rejected, then the question of Humean supervenience lapses. I conclude with a brief discussion of this position.

I. Formulating and Fixing HS

Call a property "Humean" if its instantiation requires no more than a spatiotemporal point and its instantiation at that point has no *metaphysical* implications concerning the instantiations of fundamental properties elsewhere and elsewhen. Lewis's examples of Humean properties are the values of electromagnetic and gravitational fields and the presence or absence of a material particle at a point.[3] HS says that every contingent property instantiation at our world holds *in virtue* of the instantiation of Humean properties. If *M* is a contingent property, then an instantiation of *M* holds *in virtue of* instantiations of P_1, P_2, . . . , P_n only if in every metaphysically possible world at which the *P* instantiations hold the *M* instantiation also holds.[4] Lewis illustrates the *in virtue of* relation with the example of a grid of pixels. The grid exemplifies a particular picture, say, a depiction of a cube, in virtue of the firing of the grid's pixels. Note that HS doesn't require that if an individual instantiates a property *F*, it does so in virtue of Humean properties instantiated only at points where that individual is located. The instantiation of *F* may supervene on a larger pattern of Humean property instantiations and even on their totality.

Lewis says that HS is contingent. There are un-Humean possible worlds that contain facts that fail to supervene on the mosaic of Humean property instantiations at those worlds. At an un-Humean world, for example, consciousness might be instantiated by nothing smaller than a complex organism and the totality of Humean property instantiations at that world may not be metaphysically sufficient for its instantiation. At such a world, consciousness is an *emergent* un-Humean property. HS says that the actual world contains no properties like that.

Why believe Humean Supervenience? Hume can be interpreted as advocating HS but in a very different form and for very different reasons than

Lewis. For Hume the fundamental properties are kinds of impressions instantiated in the sensorium. All true judgments supervene on the distribution of these properties. So judgments that one impression or kind of impression is nomically connected with another either are strictly false or must be construed as supervening on the distribution of fundamental properties of impressions. If this interpretation of Hume is accurate, then his version of HS is not at all plausible. But Lewis's reasons for defending HS are not Hume's. He says that he defends HS "to resist philosophical arguments that there are more things in heaven and earth than physics has dreamt of."[5] In other words, his motivation is to support physicalism. He characterizes physicalism this way:

> [M]aterialist supervenience means that for anything mental there are physical conditions that would be sufficient for its presence, and physical conditions that would be sufficient for its absence.[6]

Physicalism says that whatever happens in our world happens in virtue of physical happenings. Lewis thinks that it is the job of physics to provide an inventory of the fundamental physical properties and optimistically suggests that "present day physics goes a long way toward a correct and complete inventory."[7] Among the properties he mentions are mass and charge; properties that he also takes to be Humean. Despite this guidance he doesn't tell us what makes a property a fundamental physical one, and without such an account physicalism is threatened with vacuity.[8] A proposal that I think captures what many have on their minds when they speak of fundamental physical properties is that they are the properties expressed by simple predicates of the true comprehensive fundamental physical theory. The true comprehensive fundamental physical theory is the minimal theory that accounts for changes of the locations and motions of macroscopic spatiotemporal entities and also for changes in the properties that account for locations and motions and so on.[9] If current physics is on the right track, then charge and mass may be *fundamental* physical properties but mental properties are not. Although mental properties are invoked to account for the motions of macroscopic entities (e.g., Smith's believing that his friend is across the street is invoked to account for his crossing the street), they are not expressed by predicates of the minimal comprehensive theory that can in principle account for the motions of macroscopic entities.[10] If physicalism is true, then mental properties and all other contingent properties are instantiated in virtue of the instantiations of fundamental physical properties.[11]

HS and physicalism are distinct doctrines. HS doesn't entail physicalism since it is compatible with there being Humean properties that are not physical.[12] Physicalism doesn't entail HS since there is no guarantee that the fundamental properties posited by physics are intrinsic properties of spatiotemporal locations. In fact, it seems pretty clear that contemporary physics does dream of non-Humean properties. I have in mind so called "entangled states" that are responsible for quantum nonlocality, i.e., for quantum theory's violations of Bell inequalities.[13] The entangled state of a pair of particles fails to supervene on the intrinsic properties of the separate particles. That is, the local properties of each particle separately do not determine the full quantum state and, specifically, do not determine how the evolutions of the particles are linked.[14] Since we have reason to believe that quantum theory is true, we have reason to think that HS is false.

Lewis is aware of the objection. He initially responded by pointing out that quantum theory is not in very good philosophical shape.

> I am not ready to take lessons in ontology from quantum physics as it now is. First I must see how it looks when it is purified of instrumentalist frivolity . . . of doublethinking deviant logic . . . and—most of all—when it is purified of supernatural tales about the power of the observant mind to make things jump.[15]

However, there are versions of quantum mechanics—David Bohm's version for one—that are so purified.[16] Bohm's theory has a realist interpretation, conforms to standard logic, has no jumps at all, and doesn't figure the observant mind in its fundamental laws. The defender of HS cannot hide behind the hope that quantum theory is incoherent or merely an instrument for predicting experimental outcomes.

More recently, Lewis has accepted that HS needs to be reformulated to accommodate quantum nonlocality.[17] Here is a suggestion for how to do so in the context of Bohm's theory. The quantum state of an n-particle system is a field in $3n$ dimension configuration space where the value of the field at a point in configuration space is the amplitude of the quantum state at that point.[18] These field values can be thought of as intrinsic properties of points.[19] The ontology of Bohm's theory also includes a "world particle" whose location and motion in configuration space determines the locations and motions of ordinary material particles in three-dimensional space, and the locations and motions of these particles determine the manifest world.[20] If Bohm's theory is the correct and complete physical theory, and if physicalism is true, then everything would supervene on the quantum state and the location of the world particle. We can think of the manifest world—the world of macroscopic objects and their motions—as shadows cast by the quantum state and the world particle as they evolve in configuration space.[21]

The lesson for a defender of HS to take from quantum mechanics is to count a property as Humean in a world iff it is an intrinsic quality of points in the fundamental space of that world. If Bohm's theory (or any other version of quantum mechanics that construes the wave function realistically) is correct, then that space is configuration space.[22] Given this account of Humean properties, quantum nonlocality poses no threat to HS.

I am not sure whether my reformulation of HS takes care of all incompatibilities between HS and physics. But since Lewis's aim is to defend HS from philosophy and not from physics, let us turn to philosophical challenges to HS. The most important challenge is from philosophical views concerning nomic concepts; that is, from views about laws, causation, and chance.[23] Nomic features are not intrinsic to space-time points, so HS requires that they supervene on Humean properties. But according to the non-Humean tradition, they don't. The failure of supervenience is expressed in metaphors associated with laws and causation. Some advocates of the non-Humean tradition say that laws *govern* or *guide* the evolution of events and that causation *provides the cement* of the universe. What determines and cements *E*s can't supervene on *E*s. According to non-Humeans, the nomic facts, rather than being determined by the Humean facts, determine them! Since Lewis says that he defends HS to resist philosophical arguments that there is more than physics tells us, he must think that physics does not tell us that there are non-Humean laws or causation. But physics does not speak unambiguously. Certain regularities—e.g., Schrödinger's equation—are said to express laws, some laws posit probabilities, and physicists often claim that one event causes another (e.g., the absorption of a photon causes a change in the energy of an electron). The question is whether the "laws," "probability," and "causation" that physics speaks of are non-Humean or can be accommodated by HS. Of course, one way of defending HS is to deny that there are nomic facts. Perhaps they are projections of the mind, as Hume is reputed to have thought, or the inventions of philosophical misinterpretations of science, as van Fraassen suggests.[24] But defending nomic nonfactualism would be a Herculean undertaking. Laws and chances obviously play important roles in the sciences, and many of our concepts, for example, functional concepts, have nomic commitments.[25] So if nomic concepts are not completely factual, then the thought that a certain functional concept is instantiated is also not completely factual (or is false). Defending HS by denying nomic facts is too costly. The other way of defending HS is to show that, appearances to the contrary, nomic facts do supervene on Humean facts. More specifically, the nomic notions employed by physics and the other sciences are compatible with HS. Of course, this approach must deflate the governing and cementing metaphors that are associated with nomic concepts. But that may not be too high a price to pay if the resulting notions can do the work required of them in the sciences.

II. Lewis's Accounts of the Nomic

Lewis defends HS by constructing reductive Humean accounts of laws, chances, and causation. I will mainly be concerned with his account of laws, but it will be useful to quickly sketch his accounts of all three kinds of nomic facts. Lewis accounts for laws and chances together by building on a suggestion of Ramsey's.

> Take all deductive systems whose theorems are true. Some are simpler, better systematized than others. Some are stronger, more informative than others. These virtues compete: An uninformative system can be very simple, an unsystematized compendium of miscellaneous information can be very informative. The best system is the one that strikes as good a balance as truth will allow between simplicity and strength. How good a balance that is will depend on how kind nature is. A regularity is a law iff it is a theorem of the best system.[26]

Chances enter the picture by letting deductive systems include sentences that specify the chances of events.

> Consider deductive systems that pertain not only to what happens in history, but also to what the chances are of various outcomes in various situations—for instance the decay probabilities for atoms of various isotopes. Require these systems to be true in what they say about history. . . . Require also that these systems aren't in the business of guessing the outcomes of

what, by their own lights, are chance events; they never say that A without also saying that A never had any chance of not coming about.[27]

As Lewis says, axiom systems more or less fit the facts, are more or less strong, and are more or less simple. Strength is measured in terms of the informativeness of the implications of the axioms, fit in terms of the chance of the actual history, and simplicity in terms of syntactical and mathematical complexity and the number of independent assumptions. These features of deductive systems trade off. Strength and fit can often be improved at the cost of simplicity and vice versa. By assigning probabilities to types of events, systems sacrifice strength for fit, but they may also make great gains in simplicity. The best system is the one that gets the best balance of the three, while not both implying that q and that the chance that q is less than 1. The laws of a world are the generalizations that are entailed by the best system for that world. Among the laws may be regularities that mention chances; e.g., for any tritium atom the chance of its decaying in time interval t is x. The totality of chance laws entails what Lewis calls "history-to-chance conditionals"; i.e., statements of the form $H_t \rightarrow P_t(E) = x$. These specify the chances of future courses of events after t if the history up through t is H_t. The chance of E at t is derived from the history up to t and the history-to-chance conditionals. So as not to prejudice our discussion I will call the laws and chances delivered by this account the L-laws and L-chances. Of course, Lewis thinks that the L-laws and L-chances are the laws and chances.

A proposition is L-physically necessary just in case it is true in every world compatible with the L-laws. L-physical necessity thus defined is less than metaphysical necessity, more than mere actuality (not every truth is physically necessary), but thoroughly grounded in actuality. An interesting consequence of Lewis's account is that there are physically possible propositions that are incompatible with the laws being the laws and incompatible with their chances. We will return to this point later.

Lewis's account of causation is in terms of counterfactuals. The counterfactual $A \rightarrow B$ is true just in case there are worlds at which A and B are true that are more similar to the actual world than is any world at which A is true and B is not true. For the case in which the laws are deterministic (the indeterministic case is a bit more complicated), similarity is evaluated in terms of the extent to which worlds match the actual world in particular fact and the extent to which the spatiotemporal regions of those worlds are compatible with the laws of the actual world. Similarity in these two respects trades off. Generally, perfect match can be improved at the expense of more extensive violation of law and vice versa. According to Lewis, his account makes the counterfactual "if Nixon had pushed the button, there would have been a nuclear holocaust" come out as true. There is a world whose events conform to the actual laws and match the events of the actual world up until time *t*, when events in a small spatiotemporal region (in Nixon's brain) violate the actual laws and lead, in conforming with the actual laws, to Nixon's pushing the button a few moments later and then to the nuclear holocaust. (Of course this assumes that the button is connected, the missiles prepared, and so forth. If these conditions were not present, then the counterfactual would be false.) Lewis thinks that this world is more similar to the actual world in match and conformity to the laws than is any world at which Nixon pushes the button and there is no nuclear holocaust. Match with the actual world after the button is pushed can be restored, but only by eradicating all traces of Nixon's button pushing. Lewis thinks that this would require widespread and big violations of the actual laws.[28]

To a first approximation, Lewis's account of L-causation is: Event *e* L-causally depends on event *c* just in case *c* and *e* are distinct occurring events and if *c* had not occurred, *e* would not have occurred (or the chance of *e* would have been smaller). Event *c* L-causes event *e* just in case there is a chain of events *c* . . . *e* related by causal dependence. Of course, Lewis claims that L-causation is causation.[29]

III. Some Clarifications

Lewis's reductions of laws, chance, and causation to Humean concepts are a philosophical tour de force. If correct, they show that the nomic features of the world are compatible with HS, and that goes a long way toward demystifying them. But are his reductions correct? Like any reductions they should be evaluated in terms of how well they ground and illuminate the practices involving the concepts. These practices are reflected in and are to an extent codified by our beliefs involving them. So we need to examine whether Lewis's reductions preserve our central and supportable nomic beliefs and how well they fit in with our other well-supported views. For example, it is generally believed that laws play a central role in explanations. If this is so, then it counts in favor of the reduction of laws to L-laws if L-laws play that role, and it counts against the reduction if they don't. If L-laws (or any other nomic concepts) satisfy a sufficient number of our central and well-supported nomic beliefs and nothing else satisfies them equally well or better, then the reduction of laws to L-laws is successful. Exactly how many or which of our nomic beliefs must be respected is not clear cut. What one philosopher sees as a reduction, another may see as an elimination.[30] But if it can be shown that L-laws satisfy enough of our central beliefs concerning laws (and other nomic concepts) to play the roles that laws are supposed to play in the sciences and that nothing else plays these roles any better, then we will have good reasons to call L-laws "laws."

Since Lewis's accounts of chance, counterfactuals, and causation all involve laws, if the HS account of laws is not defensible, then even if the other accounts are correct, they would not establish that these nomic features are compatible with HS. For this reason, I will focus on Lewis's proposal that laws are L-laws.

There are some aspects of Lewis's account of laws that I want to clarify prior to seeing whether L-laws can play the role that laws are supposed to play.

Philosophers have understood "is a law" as applying to a number of different kinds of entities: sentences, propositions, or certain nonrepresentational features of reality, i.e., whatever it is that makes a particular sentence or proposition express a law. I will understand the L-laws as being propositions. They are the propositions expressed by the generalizations that are implied by the best axiom system.

"It is an L-law that *p*" is true at a world iff there is a unique best axiom system Φ for that world and among the theorems of Φ is a sentence that expresses the same proposition as "*p*." These truth conditions have some important consequences: First, "it is a law that *p*" implies "*p*." Second, "it is a law that" creates intensional contexts. So it may be a law that *F*s are followed by *G*s and it may be that *F* and *F** are coextensional, while it is not a law that *F**s are followed by *G*s. Third, what makes a proposition an L-law at a world *w* is the "vast mosaic of local matters of particular fact" at *w*. No part of that reality that can be isolated makes a general proposition lawful or accidental.

Each of the notions "simple," "informativeness," and "best" needs clarification. Lewis thinks of simplicity as an objective property of expressions in a language (e.g., a conjunction is less simple than its conjunct) or of the proposition expressed by a sentence. Some mathematical propositions are objectively simpler than others. The informativeness of a sentence is measured in terms of the number of possibilities it excludes. Lewis seems to think that the informativeness of a system is the informativeness of the conjunction of its axioms. I make a different suggestion below. Lewis doesn't say what "best" is, but it is reasonable to think of its content as being determined by scientific practice. He readily admits that all these notions are vague. But he thinks that it is not implausible that, given the way our world is, all the ways of clarifying them will count the same generalizations as laws.

There is a problem concerning the languages in which best systems are formulated. Simplicity, being partly syntactical, is sensitive to the language in which a theory is formulated, and so different choices of simple predicates can lead to different verdicts concerning simplicity. A language that contains "grue" and "bleen" as *simple* predicates but not "green" will count "All emeralds are green" as more complex than will a language that contains "green" as a simple predicate. More worrying: Let *S* be a system that entails all the truths at our world, and let *F* be a predicate that applies to all and only things at worlds where *S* holds. Then $(x)Fx$ is maximally strong and very simple. It is the best system for our world. The trouble is that it entails all true regularities, and so all regularities are L-laws.

Lewis's remedy is to insist that the simple predicates of the language in which systems are formulated (and in which their simplicity is evaluated) must express *natural* properties or universals. But which are the natural properties? One suggestion for picking out natural properties is not appropriate in the present context. It is that they are the properties that appear in the laws or that possess causal powers. This doesn't work, since it would make the analysis of laws and causation circular. Lewis's view seems to be that, since it does so much useful work, we should accept the notion of a natural property as a primitive.[31] He does say that it is plausible that the simple predicates of current physics are good candidates for expressing natural properties. But how does he know that? Perhaps Lewis's account should not be faulted for relying on the notion of a natural property since every other account of laws—both Humean and non-Humean—helps itself to a distinction between properties that are fit and those that are unfit for laws. But one worries that if the notion of a natural property is simply taken as a primitive, then we will have no epistemic access to which propositions are laws.[32] The problem isn't merely that all possible *evidence* may underdetermine which propositions are laws but that even if we know all the true sentences (except sentences that say which are the natural properties) of every possible language, we still don't know which express laws until we know which predicates express natural properties.

Here is a different suggestion for specifying the language in which the axiom systems are formulated that doesn't rely on the notion of a natural property. I assume that it is the job of physics to account for the positions and motions of paradigm physical objects (planets, projectiles, particles, etc.). This being so, the proposal is that we measure the informativeness of an axiom system so that a premium is put on its informativeness concerning the positions and motions of paradigm physical objects. And further, we measure the informativeness of a system not in terms of its content (i.e., set of possible worlds excluded) but in terms of the number and variety of its theorems. Systems have infinitely many theorems, so we just can't compare systems by counting theorems. One way to deal with this difficulty is to discount the contribution of a theorem to the informativeness of the system which implies it by the length of its proof in some regimented proof system. So in evaluating the "informativeness" of a system, we enumerate its proofs by their length, award points for the informativeness of a theorem, with extra points awarded if it is about the motions of ordinary objects, and then divide by the length of the proof.

If the above account of informativeness can be worked out, then it will immediately take care of the trivialization problem. The system $(x)Fx$ would not be counted as "informative" since, although its theorem $(x)Fx$ is very informative, it has no theorems that mention the positions and motions of ordinary objects. The other worry was that systems formulated in "gruesome" languages may vie with systems formulated in our language for simplicity and strength but may entail different generalizations. But if the systems agree with respect to the number and variety of theorems which mention positions and motions, etc., then we have no reason to believe that this will be the case, and we have some reason to believe that it won't be the case. It seems likely that the gruesome system will have to be a bit more complicated to equal an ungruesome system in informativeness. And if there are gruesome and ungruesome systems that agree in both simplicity and informativeness, they still may imply exactly the same generalizations. If this is right, then we can dispense with natural properties. But there is still an oddity. If the best system formulated in our language entails that "all emeralds are green" and "all rubies are red," then the best system formulated in a language containing the simple predicates "gred" and "emerubies" will entail that "all emerubies are gred." But maybe that's not so bad since this generalization is nomologically necessary. Perhaps our being disinclined to count it as a law just reflects the bias of the language which we actually use.

IV. Are the L-Laws the Laws?

I now want to examine to what extent L-laws satisfy our central beliefs about laws. Here is a list of the most important features that laws are supposed to have:[33]

(*i*) If it's a law that *F*s are followed by *G*s, then it is true that *F*s are followed by *G*s.

(*ii*) Being a law is a mind-independent property.

(*iii*) The laws are important features of our world worth knowing.

(*iv*) It is a goal of scientific theorizing to discover laws, and we have reason to believe that some of the propositions that the fundamental sciences classify as laws are laws.

(*v*) There is a distinction between lawful generalizations and accidental generalizations.

(*vi*) There are vacuous laws.

(*vii*) Laws are contingent but ground necessities.

(*viii*) Laws support counterfactuals.

(*ix*) Laws explain.

(*x*) Laws are confirmed by their instances.

(*xi*) The success of induction depends on the existence of laws.

(*xii*) The laws govern (direct, constrain, or probabilistically guide) the evolution of events.

(*xiii*) If it is a law that *p*, and *q* is any proposition expressing boundary conditions or initial conditions relevant to the law that are co-possible with *p*, then it is possible that it is a law that *p* and *q*.

Some of these conditions come from scientific practice and others from philosophical reflection (not confined to philosophers). Some are more important than others. Any alleged account of laws that failed to ground a distinction between lawful and accidental regularities is obviously mistaken. On the other hand, an account of laws that didn't endorse the metaphor that laws govern events shouldn't be rejected on that account. The metaphor is obscure and not obviously connected with actual scientific practice.

L-laws clearly satisfy (*i*), (*v*), and (*vi*). With respect to (*v*) and (*vi*), L-laws are a big improvement on traditional regularity accounts. According to regularity accounts, a proposition is a law iff it is expressed by a true generalization whose predicates are nonpositional and projectible.[34] Vacuous generalizations are true, so all vacuous regularities composed of projectible predicates are counted as laws by the regularity account. This can be avoided by requiring that laws have instances, but that would exclude all vacuous generalizations some of which seem to be laws; e.g., the ideal gas.

Reichenbach gives the following example to illustrate the distinction between lawful and accidental generalizations.

> (U) There are no solid one ton spheres of uranium.
> (G) There are no solid one ton spheres of gold.

Reichenbach observes that (U) is a law but that (G) isn't.[35] Both of these generalizations are true and contain only nonpositional and projectible predicates, so the regularity theory can't distinguish them. But Lewis's account can. It is plausible that quantum theory together with propositions describing the nature of uranium entail (U) but not (G). So if quantum theory is part of the best theory of our world, then (U) will be a law. In fact, the reason we think that (G) is not a law is that we think that the best theory of our world is compatible with (G)'s being false. Adding (G) to fundamental physical theory would produce a stronger system but at a great cost in simplicity.

L-laws also seem to satisfy (*iii*) and (*iv*). If one knows the L-laws, then one would know a lot about the world and have that knowledge in the form of simple compact axioms. Further, it is not implausible that, at least in physics, the goal of theory construction is to find true, strong, well-fitting, and simple theories. The fundamental theories of physics—e.g., quantum theory, general relativity—exhibit these virtues. Propositions that scientists call "laws" are consequences of the fundamental theories (e.g., Schrödinger's law) or of these laws together with sentences connecting higher-level descriptions with quantum mechanical descriptions (e.g., laws of chemical bonding). If we were to learn that a certain system was best for our world, we would have reason to believe that its general consequences are laws.

Whether or not L-laws satisfy (*vii*), (*viii*), and (*ix*) is controversial. L-laws are related to the other L-nomic concepts in more or less the way endorsed by philosophical tradition. L-laws are contingent and the regularities they entail are L-necessary; L-laws can be premises in deductive arguments that have the form of deductive nomological explanations; and it is generally the case that if it is an L-law that *F*s are followed by *G*s, then the counterfactual "if an *F* occurred, it would be followed by a *G*" will generally be true.[36] Of course, to anti-Humeans, L-laws are sham laws that are capable only of supporting sham counterfactuals, etc.[37] But these complaints should not be taken seriously unless backed up by arguments that show that L-counterfactuals, L-necessity, and L-explanation are not the genuine items. If, for example, genuine counterfactuals do not supervene on Humean facts, then they can't be supported by L-laws. But, although specifics of Lewis's account have been criticized, I know of no argument that shows that the counterfactuals laws are supposed to support express non-HS facts.

Armstrong does argue that Humean regularities cannot really explain.

> Suppose, however, that laws are mere regularities. We are then trying to explain the fact that all observed *F*s are *G*s by appealing to the hypothesis that all *F*s are *G*s. Could this hypothesis serve as an explanation? It does not seem that it could. That all *F*s are *G*s is a complex state of affairs which is in part *constituted* by the fact that all observed *F*s are *G*s. "All *F*s are *G*s" can even be rewritten as "All observed *F*s are *G*s and all unobserved *F*s are *G*s." As a result, trying to explain why all observed *F*s are *G*s by postulating that all *F*s are *G*s is a case of trying to explain something by appealing to a state of affairs part of which is the thing to be explained.[38]

It is likely that he would similarly complain that L-laws don't really explain since the fact that a regularity is an L-law is a complex state of affairs constituted in part by the regularity. But the argument isn't any good. If laws explain by logically implying an explanandum—as the DN model claims—then the state of affairs expressed by the law will in part be constituted by the state of affairs expressed by the explanandum. How else could the logical implication obtain? In any case, L-laws do explain. They explain by unifying. To say that a regularity is an L-law is to say that it *can* be derived from the best system of the world. But this entails that it can be unified by connecting it to the other regularities implied by the best system. I suspect that Armstrong thinks that L-laws don't explain because he thinks that laws explain in some way other than by unifying. I will return to this point later when we discuss his own view of laws.

Can L-laws play the roles in induction that laws are supposed to play? One of these roles is that laws are confirmed by their instances. Let's say that a generalization "*F*s are followed by *G*s" is confirmed by its instances iff an instance of the generalization increases its credibility and also the credibility that unexamined *F*s are followed by *G*s. Dretske suggests that if laws are mere Humean uniformities, then they are not confirmed by their instances.[39] He seems to think that all Humean uniformities are like "all the coins in Smith's pocket are dimes," in that one instance lends no credibility to another. But, of course, there is a difference between a uniformity which is an L-law and one which is accidental. The question is whether this difference permits confirmation of the former but not the latter. On a Bayesian account of confirmation, the answer is affirmative. There are probability distributions on which Newton's gravitational law (construed as an L-law) is confirmed by its instances but "all the coins in Smith's pocket are dimes" is not confirmed by its instances. Perhaps Dretske thinks that it should follow from the nature of laws that they are confirmed by their instances. It is true that this doesn't follow on the Bayesian account of confirmation. There are probability distributions on which gruesome generalizations rather than L-laws are confirmable. But I don't consider this to be a very strong objection to L-laws since I don't see how any plausible account of laws can *guarantee* that they are confirmed by their instances.[40]

Armstrong claims that "if laws of nature are nothing but Humean uniformities, then inductive scepticism is inevitable."[41] His argument is that if the laws were Humean uniformities, then we could not explain why induction is rational (or necessarily rational), and without such an explanation inductive skepticism follows. I don't want to examine his argument in detail.[42] Suffice it to say that it depends on his claim that non-Humean laws can explain their instances while Humean uniformities cannot. We have already seen that this assumption is question begging.

If "inductive skepticism" means that it is impossible to provide a non-question-begging justification of a system of inductive inference, then I agree with Armstrong's claim that Humeanism makes inductive skepticism inevitable. That is because it is *inevitable period,* whatever laws may be. Hume conclusively showed the impossibility of a non-question-begging justification of any universal system of inductive inference. But if Armstrong means that someone who believed that laws are Humean uniformities (or that there are no non-Humean laws) is irrational in making inductive inferences, then Armstrong is pretty clearly wrong. Suppose that *D* is a scientist who assigns a probability of 1 to HS and also allocates substantial initial probability to simple and strong theories, including the true one. As she accumulates evidence she will probably (relative to her probability assignment) come to assign a high probability to the true system and to the L-laws.[43] That her decisions are based on assigning high probabilities to many true propositions is likely to make those decisions successful. It is hard to see what reason we could have for thinking that *D* is irrational.

One of the conditions on our list that L-laws seem to violate is mind independence [i.e., condition (*ii*)]. The property of being an L-law is defined in terms of standards of simplicity, strength, fit, and best combination. These standards seem to be relative to us; i.e., to our psychology and interests. My proposal for making informativeness concerning position and motion especially important also may seem to make lawhood relative to our interests. We can imagine cognitive beings whose standards and interests differ greatly from ours. So it is apparently a consequence of Lewis's account that which propositions are laws depends on mental facts about us. This smells, at least a little, like nomic idealism.

But it is not clear that being an L-law is mind dependent in any way that is troubling for the jobs that laws are required to perform in science. First, it should be noted that the mind independence of the lawful regularities themselves is completely compatible with Lewis's account. What is at issue is whether the lawfulness of those regularities is mind dependent.[44] Second, Lewis's account is compatible with the view that scientists are now mistaken concerning which generalizations are L-laws and even with the view that in the Peircian ideal scientists will be mistaken.[45] So being an L-law is compatible with fairly robust realism. Third, the extension of "is a law" at a world *w* is determined not by the standards of simplicity, etc., of the scientists (if there are any) at *w* but by the scientists at our world. This rigidifies the standards and so falsifies the counterfactual that had our standards been different so would have the laws. Fourth, Lewis observes that simplicity, strength, fit, and balance are only partly relative to us. Independently of our psychology or opinions, a linear function is simpler than a quartic function, a second-order differential equation is simpler than a third-order one, etc. So he suggests that the mosaic of Humean facts of our world may be such that the best system is robustly the best. Varying the subjective aspects of simplicity, etc., within the space left by objective criteria may leave the best system unaltered. The upshot is that although the property of being an L-law is partly constituted by psychological factors, which generalizations are the laws is mind independent. So far as I can see, the fact that the concept of laws is partly constituted by concepts involving scientists' standards does not prevent them from explaining, supporting counterfactuals, etc.

Scientists and others often talk of laws *governing* or *guiding* events; i.e., they invoke condition (*xii*). The Laplacian creation myth embodies this way of thinking. God creates the universe by creating the laws and setting the initial conditions and then lets the history evolve under the direction of the laws. Physicists do something similar, at least in thought, when they take dynamical laws, set initial conditions, and then see what consequences ensue. But what do these metaphors of governing and guiding come to? No one thinks that the laws literally govern events.[46] Nor do the laws cause the events. But whatever these metaphors come to it is clear that L-laws don't govern the evolution of events. It is more apt to say that L-laws *summarize* events.

Condition (*xiii*) is closely connected to the idea that laws govern events. If dynamical laws govern events, then any initial conditions that are compatible with the generalizations entailed by the laws can be governed by the laws. It is not surprising, then, that L-laws don't satisfy (*xiii*). John Earman, who is sympathetic to HS, provides a simple example.[47] Consider a world *w* that contains only a single particle moving at a uniform velocity. The events of this world are compatible with Newton's laws, and it further seems possible that Newton's laws are the laws that obtain at *w*. But Newton's laws are not the L-laws at *w* since they are far more complicated and no more informative than the single generalization that all particles move at a uniform velocity.

The failure of L-laws to satisfy (*xiii*) is prima facie a serious matter. Given a set of dynamical laws, physicists consider the consequences of those laws for various initial conditions. No restriction is placed on these conditions other than that they be compatible with the generalizations expressed by the laws. L-laws can be used in this way, but there may be some initial conditions which, while consistent with the generalizations, are incompatible with their being laws.

The feeling that an adequate account of laws should satisfy (*xiii*) runs deep. Michael Tooley and John Carroll describe thought experiments which evoke intuitions based on (*xiii*) and use these thought experiments to argue against HS. Here is a variant of one of Carroll's examples: Consider worlds *u* and *v* as follows. Both *u* and *v* contain *x* particles and *y* particles and Newton's laws of motion obtain in both. The difference is that in *u* it is a law that when *x* and *y* particles interact they exchange the value of some property—say spin—while in *v* it is a law that they don't exchange spins. The initial conditions of *u* and *v* make for many such interactions. These worlds differ in their Humean facts, so there is, so far, no problem for HS. But relative to each world it is possible—i.e., compatible with its laws—for there to have been initial conditions such that, had they obtained, there would have been no interactions between *x* and *y* particles. At such worlds, do the *u* law or the *v* law concerning *x* and *y* particles hold? We can conceive of both possibilities, so it seems that there are both kinds of worlds. At *u′* it's a law that *x* and *y* particles exchange spins when they interact, and at *v′* it's a law that they don't exchange spins. At *u′* but not at *v′* it's true that if an *x* and *y* particle were to interact, they would exchange spins. Since *u′* and *v′* are identical with respect to their Humean property instantiations, HS is false. Notice that (*xiii*) is invoked in the thought experiment when it is claimed that *u′* and *v′* are possibilities.

Carroll and Tooley seem to think that this kind of thought experiment is sufficient to conclusively refute HS accounts of laws. But the argument falls short of a refutation. The intuitions involved in the thought experiment are doubly suspicious. They involve possible situations that are enormously different from the actual world, and they involve scientific concepts. The assumption that such intuitions are accurate is, at best, questionable and in some cases has been outright discredited. For example, most people have the intuitions that continued application of force is required to keep a body in motion and that the heavier the object, the faster it falls. Obviously these intuitions are misguided. Why should intuitions concerning laws be more reliable?[48]

Pointing out that intuitions are not infallible is enough to show that the thought experiments aren't conclusive refutations. But, unless they can be explained away, they do count against Lewis's reduction. That is, they count against his reduction unless it can be explained why we have such intuitions even though laws fail to satisfy (*xiii*). Although any such explanation is speculative, there is a story that strikes me as plausible for how we could come to believe, mistakenly, that L-laws should satisfy (*xiii*). According to HS, nomic facts supervene on the totality of Humean facts. It will generally be the case that in regions of space-time that are small compared to the whole spatiotemporal region of the world, events that don't violate the laws also don't violate the fact that they are the laws. That is, the following condition can be satisfied by L-laws:

> (*xiii*)* Given a set of laws {*L*} similar to the actual laws and any spatiotemporal region *S* that doesn't violate {*L*}, there is a Humean possible world *u* containing a region *S** that matches *S* and such that {*L*} is the set of the L-laws of *u*.

Physicists usually consider small systems whose initial conditions are compatible with what they take to be the lawful generalizations. Because the systems are small parts of the actual world, such systems will invariably also be compatible with these generalizations being L-laws. The practice of applying the laws to small systems (compared to the totality of facts) might lead to the belief that any system—no matter how large—whose initial conditions are compatible with the lawful generalizations is also compatible with these generalizations being laws; i.e., to (*xiii*), the condition that underlies the Carroll-Tooley intuitions. But if the laws are L-laws, then this belief is mistaken. Giving it up may be giving up something that we are used to but it wouldn't have much of an effect on scientific practice.

Let's take stock. L-laws clearly satisfy conditions (*i*), (*iii*) through (*vi*), (*x*), and (*xi*). They also satisfy (*vii*), (*viii*), and (*ix*), if the relevant nomic notions are construed as the corresponding L-nomic notions. It is arguable that L-laws satisfy (*ii*). The only conditions clearly violated by L-laws are (*xii*) and (*xiii*). Condition (*xiii*) is almost satisfied, and to the extent that it is not, we can explain why it's not though we think it should be. Condition (*xii*) is obscure. If there is nothing more to it than what is expressed by (*xiii*), then L-laws satisfy it to the extent they satisfy (*xiii*). If (*xiii*) requires something more, that more has not been expressed without metaphor and has not been shown to be anything required by science. Still, it will strike many philosophers that L-laws are eviscerated versions of laws. If there existed some entity that fully satisfied (*xii*) and (*xiii*) as well as all the other conditions on laws, then these philosophers would prefer to call these items "laws." Of course, these philosophers would thereby reject HS. If they could provide good reasons to believe that there are such robust laws, then they would provide good reasons to reject HS.

V. Non-Humean Accounts of Laws

Anti-Humeans think that Humean accounts at best deliver pale imitations of real laws. They say that real laws are distinct from the facts that they explain and don't supervene on them. I will call these hypothesized entities

"A-laws" after Armstrong who is, perhaps, their most prominent and persistent advocate.[49] A-laws are claimed to satisfy all of our conditions on laws including (*xii*) and (*xiii*). If this is so and there are A-laws, then Lewis's proposed reduction of laws to L-laws should, by his own lights, be rejected, since the A-laws, by satisfying more of our beliefs concerning laws, would better deserve the title "laws." And if there are A-laws, then HS is false, since, as we have seen, satisfaction of (*xii*) entails the failure of HS.

Let's say that "*F*s are followed by *G*s" is an A-law at a world *w* just in case the generalization "*F*s are followed by *G*s" instantiates the non-Humean property *X* at *w*. The property *X* is that property which makes "*F*s are followed by *G*s" an A-law. Armstrong, Dretske, and Tooley (ADT) all think that the property of being a law can be analyzed in terms of one property necessitating another.[50] Carroll and Maudlin propose views on which the concept of lawhood is primitive.[51] They say nothing about the property *X* that makes a generalization a law. It may be simple or complex. By offering an analysis of the law-making property, the ADT account sticks out its neck and is open to some objections that do not seem applicable to the primitivist account.[52] But the problems with A-laws that I will discuss apply to both approaches.

It is in virtue of the *X* property being non-Humean that A-laws satisfy (*xiii*), since as long as the mosaic of Humean property instantiations is logically compatible with a generalization, that generalization may satisfy *X*; i.e., it may be an A-law. It is satisfaction of *X* that empowers a generalization to explain its instances, support counterfactuals, direct or guide the evolution of events, and so forth. According to the anti-Humean, A-laws can support genuine counterfactuals; i.e., counterfactuals that don't supervene on the mosaic of Humean facts. Of course, L-counterfactuals can't do that.

There are worlds in which the A-laws and the L-laws coincide. But, of course, what makes a generalization an A-law is quite different from what makes it an L-law. There are also worlds in which the A-laws and L-laws are quite different. Two worlds can be exactly alike in their Humean facts (and therefore in their L-laws) but differ radically in their A-laws. There are worlds in which none of the generalizations entailed by the best axiom systems for those worlds are A-laws but in which other complicated and isolated generalizations are A-laws. Some worlds may have no L-laws since there are no best axiom systems for those worlds, but they may have many A-laws, etc.

Some Humeans think that the metaphysics of A-laws is incoherent. I partially agree. I don't think that there is a satisfactory way of cashing out the idea that A-laws guide or direct the evolution of events. These metaphors are supposed to provide a way of understanding how it is that laws ground necessary connections, support counterfactuals, explain their instances, and so on. For example, if we think of a law as literally directing or guiding the course of events, then it may seem that the law together with initial conditions can account for the evolution of events. For laws to operate in this way there must be a law-making feature *M* distinct from the generalization that *F*s are followed by *G*s that brings it about that *F*s are followed by *G*s. What could this bringing about be? One suggestion is that *M* *causes* *F*s to be followed by *G*s. But this is unsatisfactory. Not only do we have no idea of what this *M* is and how it causes the regularity, but the suggestion seems to involve an infinite regress. The causal relation between *M* and the regularity is presumably backed by a law that brings it about that *M* causes *F*s to cause *G*s, etc. According to Armstrong, when "*F*s are followed by *G*s" is an A-law, then the universal *F* "brings along" the universal *G* and this bringing-along relation cannot be further explained (though it is a kind of causal relation). He says that "we must admit it in the spirit of natural piety."[53]

Carroll and Maudlin drop the metaphors of directing and guiding and simply maintain that laws fail to supervene on the Humean facts. So far as I can see, there is no incoherence in their position. There are possible worlds in which some regularities instantiate a non-Humean property *X* and in which these regularities satisfy all of the conditions on laws with the exception of (*xii*). However, there are still metaphysical puzzles about A-laws. It is the fact that a generalization instantiates property *X* that is supposed to empower it to explain its instances, support counterfactuals, etc., i.e., it is that fact which makes it a law. The metaphors of directing and guiding or Armstrong's invocation of necessitation are supposed to provide some sort of an account of how A-laws explain their instances, support counterfactuals, etc. But once these metaphors are rejected it is unclear why or how the satisfaction of *X* enables a generalization to perform these feats. Carroll and Maudlin simply accept that it is a basic fact that A-laws explain, etc., without providing any account of what it is about them that enables them to do so. Their attitude is hardly different from Armstrong's recommendation of natural piety. Our reasons for believing that there are A-laws have to be very strong to justify such devotion.

So what are the reasons for believing that there are A-laws? One way of arguing for A-laws is to argue that there are laws and that L-laws (or other

Humean laws) can't do what laws are supposed to do; e.g., provide explanations, support counterfactuals, ground induction, etc. We have already discussed these arguments and found them to be question begging. There is another line of reasoning suggested by Armstrong to the effect that the existence of A-laws best explains certain regularities, and so, by inference to the best explanation, we have good reason to believe that there are A-laws.

> Laws, however, explain regularities. Even if we take the Humean uniformity itself, that all *F*s are *G*s, it seems to be an explanation of this uniformity that it is a law that *F*s are *G*s. But, given the Regularity theory, this would involve using the law to explain itself. We need to put some "distance" between the law and its manifestation if the law is to explain the manifestation.[54]

This suggests the following argument for A-laws: It is a regularity that *F*s are *G*s. There being an A-law that *F*s are *G*s explains this regularity better than its being an L-law that *F*s are *G*s. So it is reasonable to believe there is an A-law that *F*s are *G*s. There is much wrong with this argument. First, even if the existence of an A-law explained the regularity better than any competing explanation, it wouldn't follow that it is reasonable to believe that *F*s are *G*s is an A-law. At best that would make it prima facie reasonable to believe that it is an A-law. Countervailing reasons might make it unreasonable to believe that the A-law exists.[55] Second, it is not even clear that the fact that a regularity is an A-law is the best explanation of the regularity. As I previously pointed out, A-laws are simply declared to explain by postulation. In contrast, it is clear how L-laws explain. They explain by unifying. If it is an L-law that *F*s are *G*s, then the best system implies that *F*s are *G*s. Deriving *F*s are *G*s from the best system explains this regularity by unifying it.

Sometimes Armstrong suggests that our reasons for believing in A-laws are like our reasons for believing in theoretical entities. For example, we believe that electrons exist because their existence is a component of causal explanations of various phenomena; e.g., chemical bonding. It is reasonable to believe they exist because the theory that posits them unifies phenomena and provides causal explanations. But positing A-laws provides no such explanatory advantages. The hypothesis that there are A-laws which back certain regularities doesn't provide any additional unification. If anything, it is disunifying. Unlike electrons, A-laws, presumably, don't figure in causal explanations. Positing that certain regularities instantiate *X* as a theoretical explanation is doing science not philosophy. And it is doing science very badly, since it adds nothing to our scientific understanding. I conclude that these arguments that A-laws provide better or even good explanations are ineffective.

John Carroll gives an argument that can be understood as an argument for the existence of A-laws.

> [I]n order for believing or reasoning to be instantiated at least some nomic concepts must also be instantiated. So, granting that the instantiation of any nomic concept requires there to be at least one law, for the error theorist or anyone else to believe any proposition at all, there must be at least one law. Thus, like anyone else, the error theorist cannot correctly believe that our universe is lawless.[56]

I agree that this argument establishes that if anyone believes that there are no laws, then that belief must be mistaken, since belief is a nomic property. But it would be a mistake to think that it establishes that believing that there are no A-laws is pragmatically inconsistent. That follows only if the laws that are required for beliefs are A-laws.[57] But as far as I can see, being a belief can be characterized in terms of L-nomic concepts.

Here is a third argument for A-laws.

> We have reason to believe that there are laws. Furthermore, we find ourselves believing or intuiting that laws satisfy all the conditions on the list. So we have reason to believe that there are laws that satisfy all the conditions on the list. But L-laws fail to satisfy conditions (*xii*) and (*xiii*), while A-laws satisfy these conditions. So we have reason to believe that there are A-laws.

This argument involves an inference from the fact that we have certain intuitions concerning a concept *C* to the conclusion that these intuitions are satisfied by *C*'s reference. There is a long tradition in philosophy of evoking intuitions that are associated with concepts in order to discover the nature of the concepts' reference. This method seems more appropriate for some concepts than for others. But, as was mentioned earlier when discussing conditions (*xii*) and (*xiii*), when *C*'s subject matter is scientific and when the intuitions concern modality, the argument is very weak and easily defeated by alternative explanations of why we have the intuitions we do.

The arguments just canvassed provide very little reason to believe that there are A-laws. Of course, it doesn't follow that there are no A-laws, but the epistemological position of the believer in A-laws is not very attractive. The anti-Humean claims that there is a property *X* that is instantiated by certain generalizations and that it is that property which makes those generalizations genuine laws and so capable of explaining their instances. But she has no account of how *X* accomplishes this. The Humean also thinks that there is a property—being entailed by the best system—that makes a generalization a law, and she does have an account of how that property makes the generalization explanatory. Chalk one up for the Humean. Further, a traditional epistemological principle—one which is part and parcel of scientific method—is that one should not believe that a certain kind of entity exists unless that entity is required by the best explanation of accepted facts. The only "evidence" that the anti-Humean can point to that would, without begging the question, count in favor of the existence of A-laws is our intuitions of nonsupervenience. That "evidence" is very weak and can be accounted for by the Humean. Since the Humean can account for all the evidence that the anti-Humean can account for and can do so without positing non-Humean properties or anything else that the anti-Humean doesn't already accept, the epistemological principle delivers a verdict in favor of the Humean view. I think that it may be this line of reasoning that Lewis has in mind when he says that he defends HS "to resist philosophical arguments that there is more in heaven and earth than physics has dreamt of." There is no scientific reason for believing in A-laws. Of course, physics tells us that there are laws (e.g., Schrödinger's law), but it doesn't tell us whether or not laws supervene on facts. The philosophical arguments that they don't supervene depend on taking intuitions about laws much more seriously than they deserve to be taken.

VI. Conclusion

It appears, then, that L-laws are pretty good candidates for laws and that on balance we don't have reason to think that there are any better competitors. They deserve the title "laws." Does this mean that HS has been saved from philosophy? Not by a long shot. There is still chance and causation to deal with. I will just register my opinion here that a good case (very similar to the case made for laws) can be made that L-chance can play the role of chance in science.[58] I am much less sanguine about the reduction of causation to L-causation. Lewis's account of causation is bedeviled by problems involving preemption and runs into difficulties when extended to indeterministic worlds. Of course, even if L-causation isn't causation, some other HS account may work. But if no HS account of causation is correct, the situation would be very dicey. On the one hand, causality seems to be so intertwined with so many of our concepts (indeed, with the concept *concept*) that if it fails to refer, then most of our thoughts would also fail to refer. On the other hand, look as hard as one might, we just don't find causal relations among the fundamental properties of physics or in the dynamical laws of physics. So we would be in a dilemma of either rejecting aspects of our conceptual scheme or rejecting physicalism, at least in its Humean formulation.

The other philosophical threat to HS, in my view a very serious one, involves an assumption that Lewis makes concerning the relation between natural properties and laws. Interestingly, it is an assumption also made by Armstrong. The assumption is that the fundamental laws are contingent. In other words, it is metaphysically possible for a property to be involved in a law in one world but not in another. This means that a fundamental property, e.g., gravitational mass, may conform to quite different laws, or no laws at all, in different possible worlds. At first, this assumption seems plausible since laws are knowable only a posteriori. But, on second thought, the assumption that properties are individuated independently of laws is quite perplexing. It would mean, for example, that the properties of gravitational mass and positive electromagnetic charge could, in another world, exchange places in the laws of that world, or that the property of gravitational mass appears in the law $F = m_1m_2/r^5$, etc. But this seems absurd. It amounts to supposing that fundamental properties possess a kind of haeccity that makes them the properties they are independently of the laws they figure in.

The alternative, necessitarian account of laws has been around for a while.[59] Some objections to it are easy to deflect. For example, even though laws may be metaphysically necessary it doesn't follow that they are a priori or that they are necessarily instantiated. A possibly more serious objection is that if some properties are dispositional, i.e., necessarily involve laws, then others must be categorical.[60] I don't want to evaluate the viability of this view here. But it is interesting to note that if the fundamental properties are individuated by the laws in which they figure, then the debate between Lewis and Armstrong cannot get off the ground, since the issue of whether nomic facts supervene on non-nomic facts requires that we can make a distinction between the two kinds of facts. Of course, if properties are nomically individuated, then HS is false, since the instantiation of a fundamental property has metaphysical implications for the instantiations of properties elsewhere and elsewhen. The necessitarian account of laws is also at odds with the ADT account. If the fundamental properties are nomically individuated, then the laws are not, as they are on the ADT account, facts over and above occurrent events that govern or guide their evolution. Obviously, the issue of whether properties are nomically individuated needs to be settled before HS can be evaluated. But that is an issue for another paper.[61]

Notes

1. David Lewis, *Philosophical Papers,* vol. 2 (Oxford: Oxford University Press, 1986), ix.

2. E.g., Fred Dretske, "Laws of Nature," *Philosophy of Science* 44 (1977): 248–68; David Armstrong, *What Is a Law of Nature?* (Cambridge: Cambridge University Press, 1983); John Carroll, *Laws of Nature* (Cambridge: Cambridge University Press, 1994).

3. A property is intrinsic to a region *R* (or point *x*) if its instantiation at *R* doesn't metaphysically entail anything concerning contingent property instantiations at other regions. For example, being a planet is not intrinsic. Lewis's examples of Humean properties may be a bit surprising since it is natural to think that, e.g., electromagnetic field values by their very nature conform to certain laws and so have consequences for property instantiations elsewhere and elsewhen. Obviously, Lewis is not thinking of them like that but as categorical properties whose nomic commitments are contingent. I briefly discuss the plausibility of this view at the conclusion of the paper.

4. Perhaps more than this is required for one property to be instantiated in virtue of the instantiations of others. The relation expresses the idea that the first property instantiation is completely constituted by the other property instantiations. For a discussion of attempts to clarify *in virtue of,* see S. Webb, G. Witmer, and J. Yoo, "In Virtue Of" (manuscript, Rutgers University, 1996).

5. David Lewis, "Humean Supervenience Debugged," *Mind* 103 (1994): 473–89; the quotation appears on 474.

6. David Lewis, "Reduction of Mind," in Samuel Guttenplan, ed., A *Companion to the Philosophy of Mind* (Oxford: Basil Blackwell, 1994), 414.

7. Ibid., 412.

8. See the interchange between Tim Crane and D. H. Mellor ("There Is No Question of Physicalism," *Mind* 99 [1990]: 185–206) and Philip Petit ("A Definition of Physicalism," *Analysis* 53 [1993]: 213–33) for arguments that physicalism is vacuous and physicalist rejoinders.

9. Of course a fundamental physical theory—e.g., classical mechanics or quantum theory—does not by itself have any implications concerning the positions and motions of macroscopic objects. Propositions connecting the positions and motions of macroscopic objects with fundamental physical states are needed. In the case of classical mechanics, this connection is generally established by connections between macroscopic objects and the microparticles of which they are composed. The connections in quantum

theory are more complicated and controversial. (See David Albert and Barry Loewer, "Tails of Schrödinger's Cat," in Rob Clifton, ed., *Perspectives on Quantum Reality* [Boston: Kluwer, 1995].)

10. Current views might be wrong. It may be that to account for the motions of some macroscopic entities a hitherto unknown property or entity—perhaps some M-particle?—that exemplifies fundamental mental properties needs to be invoked. In that case we would say that M-particles are physical.

11. For arguments that mental properties supervene on physical properties, see Barry Loewer, "An Argument for Strong Supervenience," in Elias E. Savellos and Umit D. Yalcin, eds., *Supervenience: New Essays* (Cambridge: Cambridge University Press, 1995) and David Papineau, "Arguments for Supervenience and Physical Realization," in *Supervenience: New Essays.* Even if the physical facts metaphysically determine the mental facts (i.e., physicalism is true), it may be that we cannot epistemically determine the relation between the two. For a recent survey of attempts to explain intentional mental properties in terms of non-intentional properties, see Barry Loewer, "Naturalizing Semantics," in Crispin Wright and Robert Hale, eds., *Companion to the Philosophy of Language* (Cambridge: Cambridge University Press, 1997).

12. Nonphysical Humean properties would be epiphenomenal with respect to physical properties.

13. See Tim Maudlin, *Quantum Nonlocality and Relativity* (Oxford: Basil Blackwell, 1994) for in-depth discussion of the Bell inequalities and quantum mechanics.

14. An example of an entangled state is the EPRB state 1/2 (||>||> + ||>||>). In this state neither electron 1 nor electron 2 possesses a well-defined spin, but the state also entails that the probability of a measurement of any component of spin yielding an "up" result is 1/2. It also entails that if the spin component in the *x* direction of one electron is measured and yields "up" then a measurement of the same component of spin on the other electron will certainly yield "down."

15. Lewis, *Philosophical Papers,* vol. 2, xi.

16. In addition to Bohm's theory, two other noncollapse versions of quantum theory, the modal interpretations, and the many minds version of the many worlds interpretation are purged of the defects Lewis mentions (although each has its own peculiarities). There are also collapse versions—the GRW theory being the most promising—that are purged of the features that Lewis rightly finds unacceptable. For a survey of most of these versions, see David Albert, *Quantum Mechanics and Experience* (Cambridge, Mass.: Harvard University Press, 1992).

17. See Lewis, "Humean Supervenience Debugged."

18. This neglects spin. Particles with spin add to the dimensionality of configuration space.

19. Rendering the wave function fully Humean involves a couple of maneuvers. One is that the fact that the sum of the amplitudes of disjoint regions is less than or equal to 1 cannot be understood as following from the nature of the quantum state, since that would mean that the values of the field at some points have implications for its values at other points. Instead, this constraint has to be construed as an initial condition or law. The Schrödinger dynamics entails that if it is satisfied at one time, it is satisfied at all

times. A second problem is that the exact form of the Schrödinger equation depends on the Hamiltonian of the universe. If the Hamiltonian is understood as a property, it is not a Humean property, since it is instantiated by nothing smaller than the whole universe and doesn't supervene on Humean properties. To overcome this, it must be built into the Schrödinger law. In other words, Schrödinger's law formulated with the Hamiltonian of our universe is the fundamental dynamical law governing the evolution of the quantum field in configuration space.

20. The velocity of the world particle depends on the value of the quantum field at the point the particle occupies in accordance with a deterministic law. (See David Bohm and B. J. Hiley, *The Undivided Universe* [New York: Routledge, 1993]; Albert, op. cit.; Loewer, "An Argument for Strong Supervenience.")

21. See David Albert, "Elementary Quantum Metaphysics," in James T. Cushing, Arthur Fine, Sheldon Goldstein, eds., *Bohmian Mechanics and Quantum Theory: An Appraisal* (Dordrecht: Kluwer, 1996) for a discussion of the claim that configuration space is the fundamental space of quantum theory.

22. This version of Bohm's theory certainly isn't true. An adequate version would be one that is compatible with quantum field theory and gravitational theory. Such a theory has yet to be created.

23. Other threats come from mental properties, especially consciousness properties, and from persisting entities both physical and mental. If consciousness properties are not compatible with physicalism, then unless they were instantiated at points they would not be compatible with HS either. Whether or not instances of an *F* persist as the same *F* through time is not, at least for some *F*s (e.g., persons), obviously supervenient on Humean properties. Lewis's approach to persistence through time is to reduce it to causal relations among temporal parts.

24. Nomic nonfactualism takes two familiar forms. Noncognitivism about laws says that saying that a generalization is a law is to express an epistemic attitude toward it. Eliminativism about laws says that our concept of laws fails to refer to anything real. Blackburn defends the position that nomic concepts involve projections of our attitudes (see Simon Blackburn, *Essays in Quasi-Realism* [New York: Oxford University Press, 1993]). Bruno de Finetti advocates a noncognitivist account of chance in his "Foresight: Its Logical Laws, Its Subjective Sources," in Henry E. Kyburg, Jr., and Howard E. Smokler, eds., *Studies in Subjective Probability* (New York: Wiley, 1964). Van Fraassen argues that the place of laws within science has been greatly overestimated by philosophy (see Bas C. van Fraassen, *Laws and Symmetries* [Oxford: Oxford University Press, 1989]).

25. Carroll (op. cit.) emphasizes that nomic elimitivism is implausible for these reasons.

26. Lewis, "Humean Supervenience Debugged," 478.

27. Ibid., 480.

28. If the laws are Newton's laws or Bohm's laws (both of which are deterministic), then Lewis's account is in trouble. The problem is that because these laws are time reversible, there is a world that differs from the actual world in its history, a world in which Nixon pushes the button but a small violation of the laws of the actual world (no bigger than the violation incurred by a world that matches the actual world up until a short

time before Nixon pushes the button) leads to a world that matches the actual world from a short time after Nixon pushes the button. Because of this, "if Nixon had pushed the button, there would have been a nuclear holocaust" comes out as false on Lewis's account. One remedy is to count past match as more important than future match.

29. Lewis introduces further complications to handle probabilistic causation, preemption, and overdetermination (see Lewis, *Philosophical Papers,* vol. 2).

30. Stephen Stich, in *Deconstructing the Mind* (Cambridge, Mass.: MIT Press, 1996), discusses the difficulty of drawing a line between reduction and elimination.

31. Lewis discusses the possibility of defining natural properties as ones whose sharing makes for objective resemblance, in "New Work for the Theory of Universals," *Australasian Journal of Philosophy* 59 (1983): 5–30. But this isn't very enlightening without an account of objective resemblance.

32. Van Fraassen (op. cit.) develops this point as a criticism of Lewis's account.

33. These features are presented in van Fraassen, op. cit.

34. For sources of traditional regularity accounts, see Ernest Nagel, *The Structure of Science* (New York: Harcourt, Brace, and World, 1961); Carl Hempel, *Aspects of Scientific Explanation* (New York: The Free Press, 1965); and Nelson Goodman, *Fact, Fiction, and Forecast* (Cambridge, Mass.: Harvard University Press, 1983).

35. Cited in van Fraassen, op. cit., 27.

36. The connection between L-laws and L-counterfactuals is a bit complicated. On Lewis's account of counterfactuals, it could turn out that the most similar world or worlds in which an *F* occurs are ones in which "*F*s are followed by *G*s" is not a law and a *G* would not follow the *F*.

37. See Dretske, op. cit.; and Armstrong, op. cit.

38. Armstrong, op. cit., 40.

39. Dretske, op. cit., 258.

40. Non-Humean accounts of laws have no way of guaranteeing that scientists possess probability distributions that permit the confirmation of laws either. Of course, they could hold that confirmation is an objective notion and that scientists should have probability distributions on which laws are confirmed. But that is something Lewis could say as well. The only account of laws I know of that makes their confirmation by instances an essential feature of them is Goodman's Humean account (see Goodman op. cit.). But it fails to satisfy most of the other conditions on our list.

41. Armstrong, op. cit., 52.

42. For an examination of Armstrong's argument, see van Fraassen, op. cit., 128.

43. If, on her probability distribution, theories are underdetermined by evidence, then the best she could do (even by her own lights) is to end up allocating her probabilities among observationally equivalent theories; see John Barman, *Bayes or Bust* (Cambridge, Mass.: MIT Press, 1992).

44. Lewis jokes that "if we don't like the misfortunes of nature that the laws visit upon us, we can change the way we think!" ("Humean Supervenience Debugged," 479). Of course, at most we could change whether the misfortunes were the result of laws or were mere accidents. Changing the standards won't alleviate the misfortunes themselves.

45. Suppose that there are distinct fundamental systems that have the same observa-

tional consequences but differ with respect to some generalizations concerning unobservables. At most one is true, but even if we had all the observational evidence and followed the scientific method, it doesn't follow that we could know which is true. Hence in this situation we wouldn't know all the laws.

46. But Fred Dretske comes close.

> Consider the complex set of legal relationships defining the authority, responsibilities, and powers of the three branches of government. . . . The legal code lays down a set of relationships between the various *offices* of government, and this set of relationships imposes legal constraints. . . . Natural laws may be thought of as a set of relationships that exist between the various "offices" that objects sometimes occupy (Dretske, op. cit., 264).

47. Earman, op. cit.

48. The view that our intuitions involving a concept must be satisfied by the concept's reference may rely on a certain view of concepts and intuitions. Specifically, it may rely on the view that concepts are analytically tied to certain descriptions, that anyone who grasps a concept knows what these descriptions are, and, further, that robust intuitions involving the concept provide access to these descriptions. This view of concepts has not fared especially well in recent discussions (see Stich, op. cit.) and is, in any case, a shaky foundation on which to base a rejection of HS.

49. See Armstrong, op. cit.

50. These authors differ somewhat about the nature of the necessitation relation. Armstrong thinks of it as a causal relation.

51. See Carroll, op. cit., and Tim Maudlin, "Laws: A Modest Proposal" (manuscript, 1994).

52. For such objections, see van Fraassen, op. cit.

53. Armstrong, op. cit, 92.

54. Ibid., 41.

55. Van Fraassen (op. cit.) makes this point against "inference to the best explanation."

56. Carroll, op. cit., 91.

57. Carroll concludes from this argument that there are laws and then argues on the basis of thought experiments that laws are A-laws.

58. Lewis once thought that this was not so. (See his *Philosophical Papers,* vol. 2.) He called chance the "one big bad bug" in his HS program. The problem with L-chance is its apparent incompatibility with what he took to be the definitive principle connecting chance and belief, i.e., "the Principle Principle":

$$\text{(PP)}\ P_t(A/\text{ch}_t(A) = x \,\&\, E_t) = x,$$

where P_t is one's degree-of-belief function at time t, $\text{ch}_t(A)$ is the chance of A at t, and E_t is any proposition that is admissible at t. (PP) supplies the likelihoods in Bayes's theorem. Lewis pointed out that (PP) is incompatible with his account of chance. The trouble is this: Suppose that h is the history of the actual world up through time t, f is the actual future after t, and f^* is a non-actual future after t. The best system for $h \,\&\, f$ may endorse the history-to-chance conditional $h \rightarrow \text{ch}(f^*) = x$, even though on Lewis's account of chance the world h & f^* is incompatible with ch(f^*) = x. That is, the best system for h & f^* and h entails that ch(f^*) = x is false. This being so, the rational degree of belief to assign to f^* given that the L-chance of f is x is 0; not x, as (PP) counsels.

Lewis recently responded (in "Humean Supervenience Debugged") to his own objection to HS by adopting a suggestion due to Michael Thau to replace (PP) by

$$\text{(NP) } P(A/\text{ch}_t(A) = x \;\&\; E) = \text{ch}_t(A/B),$$

where B is the proposition that $\text{ch}_t(A) = x$ [i.e., the set of possible worlds at which $\text{ch}_t(A) = x$]. (See Michael Thau, "Undermining and Admissibility," *Mind* 103 [1994]: 495–503.) For most propositions A, the chances of A and B at t are independent, so (NP) reduces to (PP). That solves the problem.

59. See, e.g., Wilfrid Sellars, "Concepts As Involving Laws and Inconceivable without Them," *Philosophy of Science* 15 (1948): 287–315; Sydney Shoemaker, "Causality and Properties," in Peter van Inwagen, ed., *Time and Cause* (Dordrecht: D. Reidel, 1980); Chris Swoyer, "The Nature of Natural Laws," *Australasian Journal of Philosophy* 60 (1982): 203–23.

60. See Blackburn, op. cit.

61. I am grateful to David Albert, John Carroll, Brian Loar, Tim Maudlin, and Scott Sturgeon for comments on earlier versions of this paper.

11

Ceteris Paribus, There Is No Problem of Provisos

JOHN EARMAN AND JOHN ROBERTS

1. Introduction

It is often maintained that certain putative laws of nature are not strictly true unless qualified by a proviso to the effect that nothing else interferes, where what would count as an interference cannot be stated explicitly.[1] For example, consider the "law" that when the demand for a product increases while supply remains constant, the price of that product will increase. Stated thus baldly, the generalization is too strong, for there are numerous possible situations in which it would fail to obtain, such as cases of mass irrational behavior, widespread ignorance of the demand on the part of vendors, natural disasters that interfere with the normal working of the market, etc. It seems that the most we can say is that when demand increases while supply remains constant, price will increase, *unless something interferes,* i.e., "so long as other things are equal." But in this case, there seems to be little hope of finitely characterizing the class of events that would count as an interference. So it seems that our "law" is stuck with an irredeemably vague clause, something that one might have thought has no place in the statement of a law of nature. Such clauses are generally called *provisos* or *ceteris paribus clauses* in a growing literature on the topic.

The recent literature on provisos and *ceteris paribus* clauses is in agreement that such qualifications to putative laws of nature pose an important and unresolved problem.[2] There the agreement ends. Disagreement reigns with regard to the scope of the problem, its implications for the concept of a law of nature and for the status of the disciplines that employ *ceteris paribus* constructions, and the reign even extends to the very formulation of the problem. While views on the problem have proliferated rapidly, we

From *Synthese* 118 (1999): 439–78. Reprinted with kind permission of Kluwer Academic Publishers.

think that little real progress has been made toward its resolution, and this seems to us an indication that the problem has been ill-conceived; indeed, we contend that it is in need not so much of a solution as a dissolution. Since our position is bound to be controversial, we will proceed toward it by a careful plod through various attempts to deal with provisos and *ceteris paribus* laws.

In Section 2 we briefly review Lange's (1993a) attempt to state the problem of provisos as a dilemma whose horns offer either falsity or triviality. In Section 3 we review Hempel's (1988) analysis, which supposedly inspired Lange's dilemma. We argue that Hempel has been widely misunderstood; in particular, he did *not* give aid and comfort to those who claim that it is provisos all the way down to fundamental physics. At the same time, however, we argue in Section 4 that Hempel's insight (together with other plausible premises) entails that the special sciences, insofar as they remain autonomous disciplines, cannot formulate strict laws of their own. This conclusion is widely endorsed by many commentators who further conclude that the special sciences must employ *ceteris paribus* laws. They are thus obliged to confront Lange's dilemma. In Sections 5–10 we review various responses to this dilemma, all of which are found wanting. In Section 11 we argue that it is a mistake to try to provide truth conditions for *ceteris paribus* laws. When various confused and illegitimate senses of *"ceteris paribus"* are peeled away, the valid core of what is left of the problem of provisos and *ceteris paribus* clauses is a scientific, not a philosophical problem. In Section 12, we consider the nature of hypothesis-testing in the special sciences in the light of the preceding arguments. Concluding remarks are offered in Section 13.

2. An Attempt to State the Problem of Provisos

Lange (1993a) attempts to give a compact statement of the problem of provisos in the form of a dilemma which is attributed to Hempel (1988).

> For many a claim that we commonly accept as a law statement, either that claim states a relation that does not obtain, and so is false, or is shorthand for some claim that states no relation at all, and so is empty [because of the open-ended and ill-defined proviso (". . . provided that other relevant factors are absent") needed to protect the claim against counterexamples]. (235)

Applied to the "law" with which we began, the problem is clear: The statement that price *always* increases when demand rises while supply remains constant is very probably false, so we face the first horn of Lange's dilemma. In order to make the statement true, we might add the clauses, "so long as no natural disasters interfere with the market," "so long as there is no sudden outbreak of irrationality," etc. But it is clear that this won't help, because the number of interfering factors that have to be excluded is indefinitely large, and there seems to be little hope of summing them all up in a finite formulation. So we might just add the clause, "so long as nothing interferes," but then the "law" threatens to become a triviality, asserting merely that in the circumstances described, price will increase unless it doesn't. Thus, we land on the second horn of Lange's dilemma.

For the moment, let us take the dilemma at face value and ask what its scope is. It is a commonplace that the discourse in the social sciences is riddled with provisos. But Lange doesn't intend his dilemma to apply only to the social sciences; indeed, he argues that it applies equally to the hard sciences, even physics. (One of Lange's examples: the law of thermal expansion, which says that when a metal bar is heated the expansion is proportional to the temperature change, requires a proviso to ward off counterexamples such as a bar that is heated but does not expand because someone is hammering on the ends.)[3] Kincaid (1996) concurs with Lange:

> *Ceteris paribus* clauses surely do plague the social sciences. That, however, does not separate them from the natural sciences, for *ceteris paribus* clauses are endemic even in our best physics.[4] (64)

This sentiment is fairly widespread in the literature (see, for example, Carrier 1998).[5]

To get a feel for how difficult the problem is, let's do a little initial testing of the horns of the dilemma. Can we seize the first horn and maintain that genuine law statements must be true without exception or provisos and, thus, that scientists err when they attach the honorific "law" at the same time they attach a proviso or *ceteris paribus* clause? Lange thinks that the price for this move is too high. We must do justice to actual scientific practice, where proviso-ridden claims play the role of laws in that they are used to give explanations and to support counterfactuals. Should we then seize the second horn? This alternative appears even more unattractive since it seems to amount to endorsing claims of we-know-not-what.

Thus, if the advertisements are to be believed, we are faced with a pervasive problem that admits of no easy solution. We believe that there is indeed an interesting set of problems connected with provisos and *ceteris paribus* clauses. But we do not believe that anyone has succeeded in correctly diagnosing the problems. As a first step toward a diagnosis it is crucial to be clear about what the problem of provisos is *not*. Toward this end, we will review the article by Hempel that inspired Lange's dilemma. When we do that it will become clear that Hempel's problem is not Lange's dilemma.

3. What Hempel's Problem of Provisos Is Not

Hempel's (1988) discussion is couched in terms of the old fashioned "received view" of scientific theories, according to which a theory T is identified with a set of sentences which may be thought of as formulating the putative laws of the domain of the theory, as well as a set of "correspondence rules" relating terms in the observational vocabulary with the theoretical vocabulary. But Hempel gives this old view a new twist. Instead of bifurcating the non-logical vocabulary of the theory T into the observational and theoretical parts, he speaks of the antecedently understood terms (V_A) and the theoretical terms (V_C) first introduced with the theory T. (So, for example, physics might have arrived at a stage where "electron" belongs to V_A *while "quark" belongs to* V_C.) In this setting the Duhem-Quine problem amounts to the following. The idea of a hypothetico-deductive (HD) test of a theoretical hypothesis H which makes essential use of V_C, is to make predictions by deriving from H consequences stated purely in terms of V_A, and then to submit these predictions to the judgment of observation and experiment. But typically the derivation requires the help of auxiliary assumptions, with the (alleged) upshot that statements in the theory cannot be individually confirmed or disconfirmed but rather face the tribunal of experience as a corporate body.[6]

Hempel claims to have discovered a new twist to the Duhem-Quine problem: "The argument from provisos leads . . . to the stronger conclusion that even a comprehensive system of hypotheses or theoretical principles will not entail any V_A sentences because the requisite deduction is subject to provisos" (1988, 25). Hempel's claim is that typically a theory T of the advanced sciences will not have *any* logically contingent consequence S whose non-logical vocabulary belongs entirely to V_A. What we can hope to derive from T are consequences of the form $P \rightarrow S$, where again S is a logically contingent sentence whose non-logical vocabulary belongs entirely to V_A and P is a "proviso" that requires the use of V_C. If this is correct, then the instrumentalist conception of scientific theories, which views theories merely as handy devices for generating V_A predictions, is in deep trouble, a point that had already been stressed by Wilfrid Sellars (1963, 1991), for reasons similar to Hempel's.[7]

Can we express Hempel's insight without using the suspect assumption of a bifurcation of the vocabulary of theory? Following the now fashionable semantic view of theories, let us think of a theory as a family of models, and let us call those features of a model that represent phenomena that are observable independently of that theory the *empirical substructure* of that model (more or less following van Fraassen 1980). In this setting Hempel's insight, restricted to fundamental physics, amounts to the following:

> (1) For a typical theory T of fundamental physics, there are no logically contingent conditions on the empirical substructures of the models that hold across all models; but there are logically contingent conditions on empirical substructures that hold across all models in which some proviso P is true, where P places constraints on features of models other than their empirical substructures.

It should now be clear that Hempel's provisos are not Lange's provisos. Like Hempel, Lange argues for a strengthening of the Duhem-Quine problem, but his strengthening is not Hempel's. According to Lange, the auxiliary hypotheses needed to derive empirical predictions from the theory must include an indefinitely large number of presumptions, which cannot all be made explicit at once.[8] By contrast, Hempel does not suggest that it is impossible to state all of the required auxiliary hypotheses. Here it is helpful to consider Hempel's discussion of the use of Newtonian mechanics and gravitational theory to make predictions for the motions of planets of our solar system by neglecting non-gravitational forces and extra-solar system gravitational forces as well. (Assume for purposes of illustration that "force" and "mass" belong to V_C while "position" belongs to V_A.)

> [T]he envisioned application of the theory . . . presupposes a proviso to the effect that the constituent bodies of the system are subject to no forces other than their mutual gravitational attraction. This proviso precludes not only gravitational forces that might be exerted by bodies outside the solar system but also any electric, magnetic, frictional, or other forces to which the

> bodies of the system might be subject. The absence of such forces is not, of course, vouchsafed by the principles of Newton's theory, and it is for this reason that the proviso is needed. (23)

Here the proviso can be made fully explicit in a finite form. For Hempel, the important moral has nothing to do with the length of the list of the necessary auxiliary hypotheses, but rather with the fact that these hypotheses must include conditions that cannot be stated without use of the special vocabulary of the theory (here "force").[9]

More importantly, it should also be clear that Hempel's provisos are not provisos in the proper sense. By a *proviso proper* we mean a qualification without which a putative law would not be a law, not because it lacks modal force but for the more fundamental reason that it would be false unless qualified. (Recall Lange's thermal expansion example where the putative law is simply false if taken at face value without qualification.) Hempel's provisos are not provisos proper but are simply conditions of application of a theory which is intended to state lawlike generalizations that hold *without* qualification. Indeed, Hempel makes it explicit that his provisos are clauses that must be attached to *applications of a theory* rather than to law-statements,[10] in contrast to what we are calling provisos proper, which are clearly Lange's topic.[11]

This point is underscored by the fact that Hempel does entertain a doubt about whether all the provisos needed in his celestial mechanics example can be stated, but he quickly dismisses this doubt:

> The proviso must . . . imply the absence, in the case at hand, of electric, magnetic, and frictional forces; of radiation pressure; and of any telekinetic, angelic, or diabolic influences. One may well wonder whether this proviso can be expressed in the language of celestial mechanics at all, or even in the combined languages of mechanics and other physical theories. . . . It might seem, therefore, that the formulation of the proviso transcends the conceptual resources of the theory whose deductive applicability it is to secure. That, however, is not the case in the example at hand. For in Newton's second law, $f = ma$, "f" stands for the *total* force impressed on the body; and our proviso can therefore be expressed by asserting that the total force acting on each of the two bodies equals the gravitational force exerted upon it by the other body; and the latter force is determined by the law of gravitation. (Hempel 1988, 30)

Lange claims that Hempel denies that the needed proviso can be given genuine content by the theory itself, because the proviso must rule out all "other relevant factors," and no theory contains a complete list of all relevant factors (1993a, 235). But as is clear from the quoted passage, in the celestial mechanics case, Hempel takes the second law to refer to the *total* impressed force on a body, and to imply that the total impressed force (together with the mass) determines the acceleration. So a proviso to the effect that all non-negligible forces have been taken into account does imply the absence of any other relevant factors. Of course, the theory at issue may not be *true*—there may *really be* other relevant factors besides those mentioned by the theory—but what is at issue here is only whether, *given the theory,* an indefinitely large and thus unstatable host of provisos is necessary.

We fully endorse Hempel's insight. But his example and its mode of presentation are unfortunate in two respects. First, it uses an idealization (no forces acting other than gravitational forces) and/or an approximation (the total resultant forces on the planets are given to good approximation by the gravitational force component). Approximations and idealizations are widely used in physics, and their usage raises a host of important methodological issues. But Hempel's key point is independent of these issues. In the case at hand, it is in principle possible to do without any idealization or approximation: there is nothing to prevent the introduction of an explicit postulate into the theory which specifies the kinds of forces that occur in nature, and there is nothing in principle that prevents the exact specification of the values of each of these forces acting on the planets of the solar system (this specification would, of course, be a proviso, *in Hempel's sense*). Even so Hempel's key point stands: the said specification requires essential use of the V_C vocabulary. Hence, with or without idealizations and approximations, the theory by itself, without conditions of application stated in the V_C vocabulary, cannot be expected to yield non-trivial predictions stated purely in the V_A vocabulary.

Hempel's presentation appears to have misled Giere (1988), who argues that the semantic view of theories solves Hempel's problem of provisos. Giere's solution proceeds in two steps. First, take the problem of provisos to be about the role of idealizations and approximations; in particular, take it to be about (say) the role of the idealization in which there are only two bodies, the sun and the earth, acting on each other by Newton's $1/r^2$ law of gravitation. Second, claim that the problem is solved by taking Newton's laws of motion and gravitation to apply without proviso not to the messy world but to the tidy model in which there are only two bodies in the universe. But as we have urged, Hempel's key point does not concern idealizations and approximations. And even if we concentrate on the attempt to apply Newtonian theory via the indicated idealization/approximation, the semantic view hardly solves the application problem, for questions immediately arise as to the justification for using the model in question and as to how far it can be trusted to yield accurate predictions about the actual motion of the earth. Nothing in the semantic view of theories *per se* can answer these questions.[12]

The second respect in which Hempel's presentation is potentially misleading is that it has led some commentators to think that it is provisos all the way down to fundamental physics. Thus, Fodor has written that "considerations recently raised by C. G. Hempel make it seem plausible that there are *no* strict laws of nature" (1991, 21). But to repeat, the putative laws at issue in Hempel's example—Newton's second law of motion and his law of gravitation—are intended as strict laws which require no proper provisos. The notion that it is provisos all the way down to fundamental physics can be motivated by the view that the world is a messy place and that we ought not to expect to find precise, general, exceptionless laws sans proper provisos. For all we know the world could be such a messy place. Our claim is only that—contra Lange (1993), Kincaid (1996), Cartwright (1983), and Pietroski and Rey (1995)—typical theories from fundamental physics are such that *if* there they were true, there would be precise proviso-free laws. For example, Einstein's gravitational field law asserts—without equivocation, qualification, proviso, *ceteris paribus* clause—that the Ricci curvature tensor of spacetime is proportional to the total stress-energy tensor for matter-energy; the relativistic version of Maxwell's laws of electromagnetism for charge-free flat spacetime asserts—without qualification or proviso—that the curl of the **E** field is proportional to the partial time derivative of the **B** field, etc. We also claim that the history of physics and the current practice of physics reveal that it is the goal of physicists to find such strict, proviso free laws. Obviously we cannot rehearse that history here, but we believe that a fair reading of it shows that when exceptions are found to the candidates for fundamental physical laws, and when the theorists become convinced that the exceptions cannot be accommodated by explicitly formulated conditions in the language of the theory, the search is on for new candidates.

We hasten to add that what we are describing applies only to a part of physics itself and certainly not to all of physics, much less to the majority of the sciences. Indeed, we will argue below that Hempel's insight plus some other plausible assumptions make it unlikely that exceptionless laws can be formulated in phenomenological physics much less in the special sciences.[13] For us, the irony is that although Hempel's problem is not the problem of provisos (proper), his insight shows that in the broad range of cases the problem of provisos cannot be escaped. But at the same time we think it important to take a stand against the now fashionable revisionism which holds that even the most fundamental laws of physics must be qualified by provisos or *ceteris paribus* clauses.[14]

4. Hempel's Insight and the Non-Fundamental Sciences

We will now argue that if Hempel's insight is correct, then it is highly plausible that phenomenological physics, as well as the special sciences, will not be able to discover any general laws that hold without exception. Hence, if these sciences are to propose any laws at all, then these will apparently have to be *ceteris paribus* laws. (This conclusion may be thought to be obviously true. Nonetheless, we think it is interesting and worthwhile to see how its plausibility is grounded by Hempel's insight, which, we have argued, is *not* the insight that it is *ceteris paribus* all the way down.)

Phenomenological physics and the special sciences take as their subject matter entities, properties, and processes that can be observed independently of any particular theory of fundamental physics. Thus, the pronouncements of these sciences will impose conditions only on the empirical substructures of the models of any theory of fundamental physics, in the sense described above. It then follows from Hempel's insight (HI) that any generalization that these sciences discover will not be true across all models of any of our fundamental physical theories. Their truth will not be guaranteed by the laws of fundamental physics, and in that sense they will be physically contingent. Thus, if we presume that phenomenological laws or special science laws must be reducible to or supervenient upon the fundamental laws of physics, and if Hempel's insight is correct, then there are no such laws. Of course, one might well reject the presumption. Even if there are no true generalizations of, say, economics that are guaranteed to hold by the fundamental laws of physics, there might still be true lawlike generalizations about economic phenomena that have the right to be dubbed economic laws, perhaps because of the role they play in economic explanations.[15] Despite this objection, our conclusion, which we take to be a rather unsurprising corollary of Hempel's insight, enjoys widespread acceptance. In the remainder of this section, we want to illustrate how the corollary of Hempel's insight has been discovered and rediscovered, typically accompanied by great fanfare.

Consider, for example, John Beatty's (1995) evolutionary contingency thesis, according to which there are no "distinctively biological" generalizations that qualify as laws. Beatty admits that there are generalizations which apply to biological systems and which would seem to count as good candidates for laws by the usual criteria of nomicity. But he holds that they invariably fail to count as distinctively biological laws because insofar as they pass muster as laws they turn out to be wholly or largely generalizations of physics or chemistry. Beatty provides an elaborate analysis to support his thesis. We find his argument quite convincing, but, given Hempel's insight, the upshot is no surprise.[16]

Consider next Schiffer's (1991) example concerning folk psychology.

(2) If a person wants something, she'll take steps to get it.

Obviously this generalization will not stand without qualification since, for example, the person might have a stronger desire whose realization is incompatible with the realization of the first. Can all the needed qualifications be stated in the vocabulary of folk psychology or intentional psychology? It seems unlikely, for as Schiffer notes, there will probably be many nomologically possible micro-physical conditions which defeat (2) but which do not correspond to anything recognizable as an intentional psychological state.

Cartwright (1995) has discovered the corollary in economics, although she comes at it from an entirely different perspective. Her slogan is that capacities are primary and regularities are secondary.

> Fixed patterns of association among measurable quantities are a consequence of the repeated operation of factors that have stable capacities (factors of this kind are sometimes called "mechanisms") arranged in the "right" way in the "right kind" of stable environment. The image is that of a machine with set components that must be assembled and shielded and set running before any regular associations between input and output can be expected. In the case of economics we can summarize this way: *regularities*

> *are a consequence of the repeated successful operation of a socio-economic machine.* (277–78)

Adopting Cartwright's perspective we can ask: Can the conditions that capture the appropriate "shielding" of the economic machine be characterized purely in terms of economic variables (the money stock, the rate of deposits, etc.)? Cartwright thinks not (see her example on pp. 281–82). We agree. But we do not find this a surprising or profound discovery. Furthermore, adopting Hempel's insight allows us to embrace Cartwright's conclusion about special-scientific laws, while having a decent explanation of why this conclusion is true, without having to appeal to Cartwright's metaphysics of irreducible capacities (a topic to which we will return in Section 7).

5. Attempts to Cope with *Ceteris Paribus* Laws

It is generally conceded that there are no strict laws of the special sciences. A not uncommon response is that there must be *ceteris paribus* laws. Anyone who wishes to deny this response and maintain that there are no laws at all in the special sciences must cope with Pietroski and Rey's (1995) dilemma:

> [I]f one insists that the special sciences don't state laws, one must either (a) explain away the illusion that explanations like those just mentioned [using Darwin's law of fitness, Boyle's law, the law of supply and demand, etc.] avert to laws, explaining, moreover, how the special sciences can provide good explanations without having any laws to avert to, or (b) deny the immensely plausible claim that, a least sometimes, the special sciences sometimes provide good explanations. (85)

We will eventually confront this dilemma. But our initial strategy is to follow the main-line reaction in the literature to the effect that both horns are too barbed for safe engagement and that a way must be found to cope with *ceteris paribus* laws.

The ways of coping are so varied that they defy neat classification. But for present purposes we will consider them in three categories. The first and most ambitious response is to provide truth conditions for *ceteris paribus* laws, various versions of which will be discussed in Sections 7–10. A second and less ambitious response is to decline to provide truth conditions for *ceteris paribus* laws but nevertheless to show how they escape the charge of vacuity. Pietroski and Rey's (1995) version of this strategy will be taken up in Section 6. A third strategy, not unrelated to the second, is to show how *ceteris paribus* laws can be integrated with standard scientific methodology. We will briefly review two versions of this response, due to Lange (1993a) and Kincaid (1990, 1996) in the present section.

Lange's proposal is that a proviso or *ceteris paribus* clause averts to a store of implicit knowledge that is possessed by the practitioners of the science at issue and that cannot, in principle, be made explicit all at once. Lange appeals here to the Wittgensteinian point that "to require that a rule be intelligible in the absence of implicit background understanding of how to apply it is not a reasonable criterion of completeness because no rule can satisfy it" (1993a, 241). We need a rule for applying the rule, and if it is demanded that this rule be made explicit, then we still need a further rule for applying that rule, and so on. "In the same way, a law-statement specifies a determinate relation only by exploiting implicit background understanding of what it would take for nature to obey this law" (ibid.). The "rule" supplied by the thermal-expansion law is appropriately applied in some cases, not in others. In order to understand this "rule," one must know how to tell which case is at hand; for example, one must know that one ought not to apply the rule when the bar is being hammered forcefully at both ends. It isn't fair to require that all such instructions concerning when to apply the rule be made explicit, because this cannot be done for any rule. So the fact that a law is understood to have exceptions, not all of which can be made explicit, does not mean that the law is false or empty.

Let us grant for the sake of argument that the general Wittgensteinian point about rule-following is correct. The "rules" governing the use of words such as "game" or "plus" or "expansion" cannot be made fully explicit, because there is a regress of rules. You cannot explain to someone how to use these words properly unless they already have a large background knowledge of other linguistic rules that cannot all be made explicit. Still, Lange's view about laws does not immediately follow, because it is not clear that a law-statement is a "rule" in the same sense as these linguistic "rules." We take it that Newton's second law of motion is a universally quantified statement concerning the mathematical relation among the quantities mass, force, and acceleration. To explain the rules implicit in the use of the terms "mass," etc., one would have to appeal to a store of background knowledge. But Newton's second law appears to be a statement that uses these terms, rather than an attempt to explicate the rules governing their use. We cannot specify these rules in a way that is "complete" in the sense that it presupposes no implicit knowledge of other rules. But it does not follow that we cannot make a statement (such as a law-statement) that leaves no escape-clauses. This is just because we can make a statement, and make it completely (in the sense that no unstated or vaguely specified exception-clauses are needed to make it true), without stating completely all the linguistic rules that govern the terms used in the statement. For example, we can completely state the proposition that a particular apple is red, in a way that doesn't allow for exceptions and escape clauses ("What I said wasn't false, because I didn't mean that the apple was red even if someone had painted it green!"), even if we cannot specify completely all the rules that you need to be able to follow in order to understand this proposition. A law-statement has a different logical form than the statement that a particular apple is red, but it still seems to be a *statement,* so it seems that the point applies to it as well.

Perhaps this appearance is deceptive, though. Lange makes a case for the novel[17] view that, despite appearances, law-statements really are *rules,* namely rules for drawing inferences, and that as such, they are affected by the familiar regress-of-rules argument. A large part of this case is constituted by his argument that otherwise, the problem of provisos brings us to grief.[18] In response to this argument, we note first that the view that laws are rules of inference rather than statements of fact is quite counterintuitive.[19] So it seems that a rather powerful case is required to support it. But as will emerge in the rest of this paper, we don't find the threat posed by the problem of provisos to be a very strong motive, since we deny that all laws of nature are afflicted by it (see Section 3), and we think that where it does seem to pose a problem, less revisionary moves are available (see Sections 11 and 12 below).

Before moving on, we want to register one more worry we have about Lange's solution. We take it that one of the important differences between science and pseudo-science is that scientists are expected to be capable of making their presuppositions explicit, bringing them into the light of day so that they can be tested. While we do not think that there is any bright red line between science and pseudo-science, we do take it as symptomatic of the pseudo-scientific status of astrology, for example, that claims such as "*Ceteris paribus,* birth time determines personality characteristics" are defended against apparent counterevidence by the assertion that only the qualified practitioners of astrology have the tacit knowledge of when the supposed linkages do and do not hold. Lange shares this view, but we are unsatisfied by his account of the difference between science and pseudoscience: "What is noteworthy about science is that this background understanding is genuine background *understanding.* In general, all researchers identify the same testable claims as those to which one would become committed by adding a given lawlike hypothesis to a certain store of background beliefs. Because they agree on how to apply the hypothesis, it is subject to honest test" (Lange 1993a, 241–42). But it seems to us that an unspoken (and unspeakable) agreement among scientists about how to test a hypothesis does not yet guarantee that the tests are honest. Could not the scientific community as a whole capriciously and tacitly change what counts as an "interfering factor" in order to accommodate the new data as they come in (as the psychoanalytic community does, according to some critics of psychoanalysis)? This danger can be ruled out if we can say, in advance of testing, what the content of a law is, without recourse to vague escape clauses. Otherwise, we confess that we don't see how to rule the danger out. The fact (if it is a fact) that, afterwards, the scientific community generally forms a consensus about whether a rule was correctly applied, does not seem to do it. And if the danger cannot somehow be ruled out, then a proviso-ridden law-statement still threatens to become either false or trivial. So we hope that a different response to the problem of provisos is available.

We turn now to Kincaid (1990; 1996, Ch. 3) who provides an illuminating discussion of how *ceteris paribus* laws in the social sciences can explain and how they can be confirmed. In the end, however, we are left unsatisfied. Kincaid suggests that *ceteris paribus* laws are able to explain because they can pick out tendencies construed as partial causes in a causal network. When a *ceteris paribus* law takes the form, "cp: all As are Bs," there is some plausibility to the notion that it functions to indicate that A is a partial cause of B.[20] But we fail to see how this notion applies to a law of the form, "cp: ϕ" where ϕ states a quantitative functional relation. And even in cases where the tendency or partial cause notion applies, we fail to see how it underwrites explanation. For as Kincaid himself acknowledges, a tendency may be present without being dominant; and unless the tendency is dominant the actual pattern of events need not be even approximately like the pattern that would obtain if the tendency in question were the only or the dominant factor present. Thus, if what one wants explained is the *actual* pattern, how does citing a tendency—which for all one knows may or may not be dominant and, thus, by itself may or may not produce something like the actually observed pattern—serve to explain *this* pattern?

Kincaid's nine suggestions for how to confirm *ceteris paribus* laws are too complicated to summarize here. But suffice it to say that while we find much good sense in these suggestions, we are not convinced that the important problem posed by *ceteris paribus* clauses, namely the problem of their apparent lack of determinate content, has been adequately dealt with. Two of his suggestions are that (i) one can sometimes show that in some narrow range of cases the *ceteris paribus* conditions are satisfied, and (ii) one can sometimes provide inductive evidence for a *ceteris paribus* law by showing that as conditions approach those required by the *ceteris paribus* clause, the law becomes more predictively accurate. But we do not understand how to implement these suggestions unless the *ceteris paribus* conditions are known or capable of being made explicit, in which case they can be incorporated into the law and the *ceteris paribus* qualification removed. The trouble with genuine *ceteris paribus* claims is precisely that the all-things-equal clause stands for we-know-not-what and, thus, that no definite claim is in the offing. To Kincaid's suggestion that one can provide evidence that there exists some mechanism connecting the variables in the purported *ceteris paribus* law, we reply that the problem here is just a junior version of the senior problem: unless "There exists a mechanism such that . . ." can be reduced to a definite, *non-ceteris paribus* claim, the notion of evidence pro and con loses its grip.

Our discussion of Kincaid in this section may leave the impression that we object to his account of hypothesis-testing and explanation in the social sciences as such. But this is not the case; our objection is only to Kincaid's claim that *ceteris paribus* laws can play a legitimate role in scientific testing and explanation, and we think this claim can be separated out from other important claims he makes. This should become clear when we return to the issue of hypothesis-testing in Section 12, below.

6. Trying to Save *Ceteris Paribus* Laws from Vacuity

Pietroski and Rey (1995) attempt to show how *ceteris paribus* laws can be nonvacuous, without being so ambitious as to attempt to give truth conditions for such laws. They explicitly take for granted the legitimacy of a notion of scientific explanation, and a two-place relation among facts *x explains y.* They introduce the notion of *explanatory independence* $I(x, y)$ among facts as follows: $I(x, y)$ if there exists a fact z such that x explains z, z is not a logical or analytic consequence of y, and z does not depend causally on the occurrence of *y*. They consider *ceteris paribus* laws of the form:

cp: $[(x)(Fx \rightarrow (\exists y)Gy)]$

They then propose a sufficient condition for a statement of this form to be nonvacuously true.[21] Informally, their proposal amounts to the following:

> cp: $[(x)(Fx \rightarrow (\exists y Gy)]$ is nonvacuously true if each of the following three conditions obtains:
>
> (i) *F* and *G* are properties that can appear in legitimate law-statements (e.g., they are not grue-like, and perhaps they must make no essential reference to particular places, times, or objects);
>
> (ii) For every *x* such that *Fx*, there exists a *y* such that either *Gy*, or else there exists a fact *Hw* distinct from *Fx* such that *I* ([*Hw*], [~*Gy*]) and [*Hw*], either alone or in conjunction with $[(x)(Fx \rightarrow (\exists z)Gz)]$, explains [~*Gy*]. (Intuitively: For every case that fails to conform to the law, there is some fact that explains this failure (either alone or in conjunction with the law itself), where this fact does some explanatory work independent of explaining this failure. This is to rule out viciously ad hoc appeals to dubious "facts" to explain away every failure of one's favorite theory.)
>
> (iii) There is at least one concrete case of an *x*, such that *Fx*, and a *y*, where either *Fx* together with the law explains *Gy*, or there is some independent explanation of ~*Gy* as per clause (ii).[22]

The promising idea behind this account is that a *ceteris paribus* law can be nontrivially true, even if we don't cash out explicitly which "other things" need to be kept "equal," if for every occurrence of the antecedent of the law, either the consequent holds, or there is some independent interfering factor that can explain why the consequent doesn't hold. The requirement in clause (ii) that *I*([*Hw*], [~*Gy*]) is intended to require that the interfering factor [*Hw*] is indeed independent, and is not simply an ad hoc "fact" cooked up to explain a particular failure of the law.

Unfortunately, conditions (i)–(iii) are not sufficient for the nonvacuous truth of the *ceteris paribus* law. To see why, let "*Fx*" stand for "*x* is spherical," and let "*Gy*" stand for $y = x$ and *y* is electrically conductive." Now, it is highly plausible that for any body that is not electrically conductive, there is some fact about it—namely its molecular structure—that explains its non-conductivity, and that this fact also explains other facts that are logically and causally independent of its non-conductivity—e.g., some of its thermodynamic properties. Thus, clauses (ii) and (iii) appear to be easily satisfied. If Pietroski and Rey's proposal were correct, then it would follow that *ceteris paribus,* all spherical bodies conduct electricity. More generally, whenever any object's failure to exhibit property G can be explained by anything independent of whether the object exhibits property F, then Pietroski and Rey's proposal implies that *ceteris paribus,* anything with property F also has property G. Surely this trivializes the proposal, so that it does not, after all, provide a sufficient condition for the nonvacuous truth of a *ceteris paribus* law.

The general moral of this observation seems to be that it is not enough simply to require, as Pietroski and Rey do, that when cp: $(A \rightarrow B)$, any case of A accompanied by $\sim B$ must be such that there is an independent explanation of $\sim B$. This is because this requirement does not guarantee that A is in any way relevant to B, which surely must be the case if cp: $(A \rightarrow B)$ is a law of nature. Perhaps Pietroski and Rey's proposal could be modified to remedy this defect. But we do not see how to do this other than by requiring that the antecedent of the law be relevant to its consequent, in a previously understood sense of "relevant." It is not clear to us that the relevant sense of "relevant" would not depend on a previously understood concept of a *ceteris paribus* law, rendering the account circular. Of course, one could simply take the required notion of relevance as a primitive, but this strikes us as a very unattractive move, since we take it that the kind of relevance in question is something we understand by way of our notion of a law.

7. Providing Truth Conditions: Cartwright's Account

We have considered attempts to cope with *ceteris paribus* laws by means of explaining how they can play a legitimate role in scientific practice, and by showing how they can be nonvacuous, and have found these attempts wanting. Now we turn to more ambitious projects, which seek to come to terms with *ceteris paribus* laws by specifying their content or truth conditions. In this section we deal with a proposal developed in many writings by Cartwright.

Cartwright (1989, 1995, 1997) argues that the law-statements formulated by the sciences, if construed as statements of regularities in the course of events, are not true without qualification. If construed as generalizations about how empirical phenomena unfold, they must be construed as true only *ceteris paribus.* However, she holds that this way of putting the matter obscures the true role of laws in science, because she claims that law statements (and in particular, purported statements of fundamental laws) should not be interpreted as statements of regularities or generalizations about the course of events. Rather, they are attributions of capacities and tendencies[23] to various kinds of systems; in a nutshell, "cp: $(x)(Fx \rightarrow Gx)$" is true just in case all *F*s have a capacity or tendency to be *G*, so that they will be *G* in (the rare) cases where there are no other capacities or tendencies acting on them. Furthermore, she argues that such attributions do not entail any strict regularities about how empirical phenomena unfold. So statements that purport to assert lawlike regularities in the observable course of events can be true only if qualified by a *ceteris paribus* clause. This is, she claims, equally true of physics and the social sciences.[24] We will object to her argument that laws, in her sense, do not imply any regularities that hold without *ceteris paribus* qualification, and to her claim that attributions of capacities that do not imply any such regularities can be empirically confirmed and play an important role in empirical science.

For Cartwright, a typical law says that systems of kind *A* have a certain capacity *C*, and such a claim does not entail any regularities concerning the behavior of *A*s, because the way in which any particular *A* will behave depends on what other capacities it has, what capacities are possessed by the systems with which it interacts, and the ways in which all these capacities interact and interfere with one another.[25] For example, if we accept as a law the proposition that a magnet has a capacity to attract steel, nothing follows about what will happen if we place a magnet near a steel paper clip; what will happen will depend on what other factors are in play. The most that we can infer from our law is that the paper clip will be drawn to the magnet unless some other capacity interferes with the attractive capacity of the magnet in such a way as to prevent this.

But all that this argument shows is that whatever regularities we can infer from the law will have to be stated in a vocabulary that includes terms referring to other capacities. This point is very similar to Hempel's insight, which tells us that the laws of a theory will not imply any regularities that can be stated without using the vocabulary of the theory, which will include the vocabulary we need to discuss the capacities of the magnet and perhaps other capacities as well. (Indeed, Cartwright notes a strong parallel between her argument and an argument due to Sellars, which is the same argument that we have already noted is mirrored by Hempel's 1988 argument.[26]) This kind of consideration does not show that, if we avail ourselves of a rich theoretical vocabulary that allows us to refer to capacities and other theoretical items, we will still be unable to state laws that imply strict regularities governing the course of events.[27]

However, Cartwright thinks that we will not be able to state such laws, even if we allow ourselves to refer to capacities. She asserts that any real natural system will be subject to the influences of a set of capacities that cannot, in principle, be covered by any of our scientific theories, or even by all of our theories put together.[28] If we attempt to formulate a regularity that will allow us accurately to predict the behavior of a given system of a given kind in a certain set of circumstances, we can begin by enumerating all of the capacities that, according to our theories, a system of this kind possesses, as well as the capacities that, according to our theories, are possessed by the other systems with which this system interacts, and all of the laws we have on hand that concern these capacities and how they interact with one another. This process might go as follows: We start by writing down a regularity that describes how the system would behave if only one of its capacities were in operation; then, one by one, we make the corrections to this regularity that are called for by the other capacities in play and the laws concerning these capacities; but in the end, our corrected regularity will not be corrected enough, because there will always be further capacities (or perhaps other interactions among capacities), specific to the given context, that are not and in principle cannot be covered by any theory.[29]

We do not know how to begin to assess Cartwright's claim about context-specific factors that in principle elude theoretical treatment: it appears to be a very strong metaphysical thesis concerning the disorderliness of the world (perhaps motivated by her Anglocentric theological view that "God has the untidy mind of the English"[30]). But consider what would follow if it were true: none of our theories, and not even all of our theories taken together, would suffice to make a reliable prediction of any course of observable events. In fact, it appears that any course of events would be compatible with any set of laws (understood in Cartwright's sense as attributions of capacities and tendencies), for any deviation from what one might have expected given those laws could be explained away as the result of context-specific factors not captured by the net of theory. Given this, it is difficult to see how laws, as Cartwright understands them, can be used for making predictions or giving explanations, and it is far from clear how hypotheses about such laws could be confirmed. (Cartwright appeals to Glymourian bootstrap methodology as a way of showing how claims about capacities can be confirmed: we confirm new hypotheses about capacities relative to a background theory that already includes other claims about capacities. But it is hard to see how this solves the present problem; if the background theory consists of claims about capacities that entail no observable regularities, and the hypothesis to be tested is also such a claim, then how can any testable consequences be derived from the background theory together with the hypothesis?)

The arguments we have criticized have a place in a complicated and subtle view of the way that science works, and we have little doubt that Cartwright can produce an interesting response to our objections, but we do not see how a satisfactory response could go. So we tentatively conclude that the arguments of Cartwright just discussed provide neither a good reason for thinking that the laws of nature do not entail any strict regularities concerning the course of events, nor a satisfying way of understanding how "laws" that do not entail such regularities can play a role in science.

8. Providing Truth Conditions: Silverberg's Analysis

A recurring complaint against *ceteris paribus* claims is that they have "no clear meaning" (Hutchison 1938, 1965). Silverberg (1996) takes this complaint to be equivalent to the charge that *ceteris paribus* claims are semantically defective, and he seeks to rebut the charge by providing a possible-worlds semantics for them. His account makes use of David Lewis's (1973) notion of a relation, defined over possible worlds, of comparative similarity to the actual world. Whereas this relation is rarely specified explicitly, Lewis contends that in any given context, such a relation is implicitly in use. Silverberg further proposes that we distinguish between worlds that are appropriately idealized and those that are not, where what counts as an appropriately idealized world is fixed by pragmatic factors depending on the context of scientific practice in which a *ceteris paribus* law is stated. On Silverberg's account, it is a law that cp: (if *A* then *B*) just in case *B* is true in all possible *A*-worlds that are appropriately ideal and that are otherwise most similar to the actual world.[31]

Silverberg's analysis is targeted at *ceteris paribus* claims that involve idealizations. This covers many examples of *ceteris paribus* laws from economics, where appeal is made to perfectly rational agents, perfect information, perfect markets, etc. But not all of the cp laws of the special sciences are of this sort, and in general the problem of *ceteris paribus* qualifications is distinct from the problem of idealizations. Often the idealization can be stated in a precise closed form (e.g., the ideal gas law assumes that the gas molecules have no volume and interact only by contact). Here the problem is not in saying precisely what is involved in the idealization but in relating it to the real world which is not ideal. By contrast, many cp laws claim to be about unidealized real world situations but make indefinite claims about these situations. This leads to our second qualm about Silverberg's analysis. When Hutchison charges that *ceteris paribus* claims have "no clear meaning," his complaint is not that there is no respectable semantics for such claims but rather that they are pseudoclaims because they make no definite assertions and, thus, cannot be used in science to make predictions and cannot be confirmed or disconfirmed by the usual methods of scientific inquiry. Providing a possible-worlds semantics for them of Silverberg's type may provide us with a way of understanding their truth conditions, and hence how they can make a definite claim. But it does not yet show us how the definite claims they make can be tested empirically, so it seems that it doesn't show how they can play any role in empirical science.

9. Providing Truth Conditions: Fodor's Analysis

Fodor (1991) develops an analysis of the truth conditions of *ceteris paribus* laws that presupposes the notion of a law simpliciter. We think that his proposal, combined with a helpful suggestion by Silverberg (1996), is promising, despite objections raised against it by both Silverberg (1996) and Mott (1992). Fodor considers putative laws of the form, cp: $(A \rightarrow B)$, where A *is* a functional state, e.g., a psychological state, which can be realized by a number of non-functional states that may be called its *realizers.* If R is a realizer of A, then in Fodor's terminology, if a condition C is such that an occurrence of R together with C is nomologically sufficient for B, but C alone is not, then C is a *completer* for R and the law cp: $(A \rightarrow B)$. Fodor considers the proposal that cp: $(A \rightarrow B)$ is true if and only if every realizer R of A has a completer for R and cp: $(A \rightarrow B)$. But he rejects this because of an argument of Schiffer (1991) to the effect that for any psychological state A and any typical, plausible psychological law cp: $(A \rightarrow B)$, there are likely to be realizers of A for which there are no completers. So Fodor instead proposes that:

> (3) cp: $(A \rightarrow B)$ is true if (and perhaps only if): Every realizer R of A either has a completer C for cp: $(A \rightarrow B)$, or else it has a completer for sufficiently many other laws with A in their antecedents.

Against proposal (3), Mott (1992) has argued that it trivializes the notion of a *ceteris paribus* law. For example, consider the ludicrous statement that *ceteris paribus,* if a person is thirsty, then she will eat salt. It seems plausible that many (and probably all) neurophysiological realizers of the state of being thirsty will lack completers for this law. However, each of these realizers will presumably have completers for a great many other psychological laws involving thirst. Hence, this ridiculous non-law will achieve the status of a *ceteris paribus* law by piggybacking on all of the legitimate laws.

Silverberg proposes an amendment to Fodor's (3) designed to get around this objection. The amended proposal is as follows:

> (4) cp: $(A \rightarrow B)$ is true if (and perhaps only if): Every realizer R of A is such that either R has a completer C for cp: $(A \rightarrow B)$, or else both: (i) R has a completer for sufficiently many other laws with A in their antecedents, and (ii) sufficiently many other realizers of A do have completers for cp: $(A \rightarrow B)$.

This seems to evade Mott's objection. In the case of the putative law that *ceteris paribus,* thirsty people eat salt, very few (if any) realizers of the antecedent will have completers, so the putative law will not be able to piggyback its way to lawhood.

Silverberg, however, is not satisfied with (4) because he objects to the use of the phrase "sufficiently many," which he suggests is just as vague as the phrase "*ceteris paribus,*" undermining the usefulness of (4). While we share this qualm, we also think that Fodor's analysis points to an important feature of *ceteris paribus* laws. But before trying to spell this out, we will review one more attempt to provide truth conditions.

10. Providing Truth Conditions: Hausman's Analysis

Hausman (1991) maintains that "*ceteris paribus*" has an invariant meaning—namely, other things being equal—whereas the property or proposition it picks out varies with the context. He proposes that "cp: (Every F is a G)" is true in context X just in case X picks out a property C such that "Everything that is both F and C is a G" is true. Further, "cp: (Every F is a G)" expresses a law just in case the cp clause determines a property C in the given context such that "Everything that is both F and C is a G" is a law in the strict sense.[32]

The first worry we have about this proposal is that it is at once too strong and too weak. In Fodor's terminology, Hausman's *C* is a completer.[33] Schiffer's (1991) and Fodor's (1991) point, which strikes us as correct, is that for the *ceteris paribus* laws of psychology one cannot expect that there will be an appropriate completer *C* that covers every physically possible case; hence Hausman's account is too strong. But even if in the case of some particular law there is a completer for every physically possible situation, this is not very helpful from the point of view of psychology if *C* obtains only rarely among the intended applications of the (would be) psychological law; hence Hausman's account appears to be too weak, since it seems to require an additional clause to the effect that *C* obtains commonly among the intended applications.

Our second worry is more general. It concerns the question of how the context determines the content of the *ceteris paribus* clause. Hausman has little to say about this matter, but the vagueness of the proposal is not sufficient to hide the following problem. Hausman's analysis is intended to apply mainly to economics, and what he has in mind for the content of the *ceteris paribus* clause in, for example, cp: (if the price of a good falls, the demand for it will rise) in standard economic contexts is something like the condition that the prices and the tastes for other goods remain the same. But given Hempel's insight and given that the laws of economics (if there are any) supervene on the laws of physics, it is wholly implausible that the content of the *ceteris paribus* clause can be specified in purely economic terms so as to yield a strict law, as required by Hausman's analysis. And it is beyond plausibility that the economic context will pick out the fundamental physical properties needed to underwrite a strict law.

While we find Hausman's analysis of the truth conditions of *ceteris paribus* laws wanting, we find considerable merit in his discussion of the factors, such as reliability, that make it reasonable to believe that a *ceteris paribus* generalization expresses a law.[34] We will make use of these merits in our own proposal.

11. *Ceteris Paribus* Laws: The Problem That Isn't and the Problem That Is

Contemporary philosophy of science has a healthy naturalistic tenor: the job of philosophy is not to issue normative dicta to scientists but to reconstruct and clarify what actually goes on in science. It seems to follow that, to the extent that *ceteris paribus* claims are employed in some special science, it behooves the philosopher to provide an analysis on which of these claims play a respectable role in that science. While sharing the general naturalistic orientation, we think that the attitude is misapplied in the present case. In a nutshell, our position is that there is no distinctively philosophical problem about *ceteris paribus,* but there is a scientific problem: what is needed is not finer logic chopping but better science. Since this position is bound to be controversial, we will approach it indirectly through some logic chopping of our own.

Using insights gathered from the failures of the analyses reviewed above, we will propose a set of truth conditions for *ceteris paribus* laws. Let ϕ be a lawlike generalization of *X,* where *X* is some special science; that is, the non-logical vocabulary of ϕ is appropriate to *X,* and ϕ satisfies the usual criteria for lawlikeness. We seek conditions for "cp: ϕ" to be a "law of *X,*" where the scare quotes indicate that "cp: ϕ" may not be a strict law but that, nevertheless, it has features that imply that it functions like a law in *X*. Our task falls into two parts: (A) specifying the conditions for "cp: ϕ" to be true for *X,* and (B), specifying what additional conditions are needed for "cp: ϕ" to be a "law of *X*." We begin with the second subtask.

Our idea is that the additional condition needed for ϕ to be a "law of *X*" is for ϕ to have for *X* an analogue of the feature that allows us to identify something as a strict law. This is an obvious sticking point since the philosophical literature is badly at odds on how to analyze the notion of a strict law of nature. Purely for purposes of illustration, suppose that David Lewis's (1973) "best system" analysis is correct: a strict law is an axiom or a theorem of the deductive system (deductively closed and axiomatizable set of true sentences) that achieves the best balance between strength and simplicity.[35] Assuming that this is right, then for "cp: ϕ" to be a "law of *X*" we want it to be the case that there are other lawlike generalizations ϕ′, ϕ″, . . . , of *X* such that "cp: (ϕ & ϕ′ & ϕ″ & . . .)" is true for *X* and such that if all these ϕs were true and if the world were completely described in the vocabulary of *X,* then the ϕs would form the axioms of the best deductive system. If some other account of lawhood is preferred, then the story here will have to be different, but we maintain that whatever distinguishes *ceteris paribus* laws from merely contingent *ceteris paribus* generalizations is just whatever distinguishes strict laws from contingent strict generalizations.

The first subtask turns out to be more difficult since, we contend, there is no univocal sense of "cp: ϕ." The ragged character of the philosophical literature on this topic is explained in part by the fact that it tries to treat under one umbrella several different usages. For example, there is the *lazy sense* of *ceteris paribus,* as in Lange's example of "cp: (if a metal bar is heated uniformly, its expansion is directly proportional to the difference in temperature before and after heating)." We contend that when physicists assert the heat expansion law they are implicitly assuming that there are no external stresses acting on the bar. If so, this assumption can be explicitly incorporated into the generalization, obviating the need for a *ceteris paribus* qualification.[36] We assume subsequently that the *ceteris paribus* qualification is not being used in this lazy sense. We also exclude the *improper sense,* as in "cp: (any two oppositely charged particles attract each other with a force inversely proportional to the square of the distance between them)." We have maintained that "*ceteris paribus*" is out of place here because what physicists intend to assert is the unqualified strict law that two oppositely charged particles *always* exert an electrical force on one another of the form indicated.[37] We also think that it is improper to classify Boyle's law as a *ceteris paribus* law. The obvious things to say here are that it is not a law because it is false; that it is false because it is based on unrealistic idealizations (e.g., that the gas molecules have no volume); but that, nevertheless, for some gases and some pressure-temperature ranges the idealization provides an approximation that is good enough for most applications. Nothing is gained or clarified by slapping on an "others things equal" clause.

As for the proper and non-lazy senses, the literature reveals two main strands that give rise to a weak and a strong sense of "*ceteris paribus.*" We begin with the weaker one, which can be described informally as follows: "cp: ϕ" is a truth of the science X just in case there is an important class of cases in which ϕ is true, and systematic violations of ϕ cannot be produced, at least not using the techniques appropriate to X. More formally, we define this weak sense of "*ceteris paribus*" in two clauses. (i) There should be no condition Ψ which can be stated in the language of X, which may not occur "naturally" but which can be realized using the techniques of X, and which defeats ϕ. By "defeats ϕ" we mean that when Ψ obtains, ϕ is not even approximately true for X. Here we are following Mott (1992) who has emphasized that if violations of ϕ can be systematically produced, then "cp: ϕ" is not a law in any interesting sense; indeed, "cp: ϕ" is not even true. Of course, if such a factor Ψ is discovered, then ϕ could be modified to $\hat{\phi} = {\sim}\Psi \rightarrow \phi$. But this could cause trouble if the best-system analysis is accepted, since the move from a system that entails ϕ to one that entails $\hat{\phi}$ but not ϕ may represent a loss of both strength and simplicity and so $\hat{\phi}$ may not qualify as a 'law of X' on this analysis (and of course, it is possible that this move will cause problems given other analyses of lawhood as well). As for approximate truth we wish to say only that while there is no good analysis of approximate truth in general, in typical concrete cases of quantitative relations, scientists have no trouble in making precise what it means for such a relation to be approximately true.[38] The qualification "for X" in "approximately true for X" is added to emphasize that judgments of approximate truth are to be made relative to the kinds of measurements and empirical data available in X (e.g., in phenomenological physics, the data concerns macroscopically discernible states and changes of state). We are not committed to drawing unbreachable boundaries between sciences in terms of vocabulary and techniques; indeed, we think that it is conceivable for just about any science to incorporate vocabulary and techniques from other sciences. But we are committed to the view that a *ceteris paribus* law involves implicit reference to boundaries drawn in this way and, thus, that what counts as a *ceteris paribus* law for X changes as these boundaries shift.[39] Further, in the spirit of Hausman (1991), we require that (ii) there be conditions θ such that when θ obtains, ϕ is true or approximately true for X. Since we are not dealing with strict laws, we cannot argue, as we did in Section 4, that Hempel's insight implies that θ cannot be stated in the vocabulary of the special science X. Nevertheless, we think that if ϕ is not only a "law for X" but is in some appropriate sense near a strict law, then the conclusion will continue to hold.

This *weak sense* of "cp: ϕ," captured by the conjunction of (i) and (ii), is compatible with θ's obtaining only rarely in the intended applications of ϕ in X. If in addition a failure of θ to obtain defeats ϕ, then the weak sense of "cp: ϕ" has the unappetizing feature that "cp: ϕ" can be true for X even though ϕ gives (by the standards) of X, completely unreliable predictions for its intended applications. Even if θ obtains often in the intended applications of ϕ, there is little to comfort the practitioners of X if they are not in a position to determine when θ obtains and when it doesn't. And assuming the implication of Hempel's insight extends to ϕ, such a determination will involve theorizing about entities and processes that may be regarded as beyond the ken of X. If so, "cp: ϕ" degenerates into a kind of *wannabe* sense: here "*ceteris paribus*" is an implicit admission that X has not achieved reliable generalizations and, perhaps, also as an expression of a pious hope that a reliable generalization is to be found in the neighborhood.

Such reflections motivate the move to the *strong sense* of "cp: ϕ," which adds a third clause inspired by Fodor (1991) and Silverberg (1992): (iii) the condition θ of (ii) obtains in "most" of the intended applications of ϕ in X. The conjunction of clauses (i)–(iii) appears to be a good statement of what is generally intended by "*ceteris paribus*" in most of the cases considered by the philosophers we have been discussing: a *ceteris paribus* law (in the strong sense) is a generalization that plays some of the roles of laws in the science at issue, and that is not strictly true but that, nevertheless, is approximately true in most of its intended applications: extraordinary situations (e.g., for the case of psychological *ceteris paribus* laws, severe neurophysiological malfunction) may render the generalization false, but such situations are not among the intended applications, and in most of the intended applications, the generalization is reliable. But this strong sense of "*ceteris paribus*" obviously only has a determinate content when we have a reasonably clear sense not only of what "approximately true" (as used in (ii)) means, but also of what is meant by "most of the intended applications." We now wish to illustrate how determinate senses can be supplied to clauses (ii) and (iii) and how our analysis of the strong sense of "*ceteris paribus*" can be clearly satisfied.

The laws of classical phenomenological thermodynamics appear to satisfy conditions (i)–(iii). One can be confident in making pronouncements about clauses (ii) and (iii) because of the reduction of thermodynamics to statistical mechanics.[40] There are dynamically possible microtrajectories that produce permanently anti-thermodynamical behavior, but it is thought that the set of such trajectories is ignorable in the sense of being "measure zero" in the phase space of the system. It is also consistent with basic physical laws that fluctuations eventuate in temporary anti-thermodynamical behavior, but it is extremely unlikely that fluctuations will produce violations of the second law of thermodynamics, say, in the form of decreases in entropy in a closed system, which can be detected on the macroscopic scale. Violations of the second law are detectable with the aid of a low-power microscope—for example, in Brownian motion. This led to the worry that small fluctuations could somehow be exploited by clever devices so as to systematically produce macroscopic violations of the second law in the form of a perpetual motion machine of the second kind that would output macroscopically usable work. However, the long history of failed attempts along these lines, plus some theoretical considerations, strongly indicate that clause (i) is safe.[41] Furthermore, it is clear that, by the account sketched above, we are justified not only in saying that the laws of phenomenological thermodynamics are *ceteris paribus* true in the strong sense, but also that they are *ceteris paribus laws:* the historical evidence clearly indicates that the laws of thermodynamics do function as laws of phenomenological physics. Further, in the nineteenth century when the laws of thermodynamics were thought to be strictly true and when many physicists were dubious of atomism and thought that the world could be fully described in phenomenological terms, the honorific "laws" was bestowed on these generalizations.

Note the important role played here by the availability of a micro-reduction: the reduction of thermodynamics to statistical mechanics is what makes it possible to give a clear sense to the crucial phrase, "most of the intended applications." We think that this will probably turn out to be true in general. Consider again the Fodor-Silverberg proposal (4). In order to make it clear that we have a "*ceteris paribus* law" in the sense of this analysis, we would need to have at least a sketch of a micro-reduction of psychology to neuroscience, so that we can make clear what the relevant "realizers" and "completers" are. If a sketch of such a reduction is supplied, then it might well turn out that there emerges a natural measure over the microstates that play the role of realizers, so that a determinate sense can be given to the phrase "sufficiently many" as it occurs in the analysis. But the importance of reductionism is a side light to our main point, which is that in order for clauses (ii) and (iii) to make any determinate claim about the world, a determinate sense must be given to the slippery notion of "most of the intended applications."

But there is obviously something strange about our illustration: scientists generally do not attach a *ceteris paribus* clause to the laws of thermodynamics.[42] Furthermore, in cases where *ceteris paribus* clauses are typically attached (e.g., in psychology and economics), it seems impossible to satisfy clauses (ii) and (iii) with any perfectly determinate sense of "most of the intended applications." This sets up the moral we wish to draw. We claim that (1) our strong sense of "*ceteris paribus*" captures the intuitive notion that philosophers usually have in mind when discussing *ceteris paribus* laws, and (2) this sense can be made precise, in a way that makes it clear what claim about the world is made by a *ceteris paribus* law so that it is clear how such a law can be confirmed and support predictions, only if a determinate sense can be given to "most of the intended applications," but (3) the clear cases where this demand can be satisfied are cases where the phrase "*ceteris paribus*" is not used by scientists and where its use feels out of place.

There *is* a clear sense to be given to the notion of a "near-law," i.e., a generalization that is not a strict law, but that deserves to be called a "near-law" because it is, in a precise sense, true or approximately true in almost all intended applications, because it plays the role of laws in giving explanations, supporting counterfactuals, etc., and because it is clear that it makes definite claims about the world and can be confirmed or disconfirmed empirically. But, we claim, the most clear paradigms of such laws (viz. the laws of phenomenological thermodynamics) are not thought of as *ceteris paribus* laws, and statements that are thought of as *ceteris paribus* laws do not answer to this clear sense of a "near-law." Our conclusion is that the use of a *ceteris paribus* clause is a flag indicating that this kind of near-law has not been found, but that some vague but perhaps more-or-less useful generalization has been found, perhaps with the hope that a clear case of a near-law is "in the neighborhood." But since clauses (ii) and (iii) cannot be seen to be true in any determinate sense of "most of the intended applications," such generalizations do not make definite claims about the world, and so, we maintain, it is hard to see how they can be empirically confirmed or disconfirmed and what role they can play in making scientific predictions and giving scientific explanations.

In the light of this, we wish to make the following suggestion. "*Ceteris paribus* laws" are not what many philosophers have taken them to be; that is, they are not elements of typical scientific theories that play the same kinds of roles in the practice of science that less problematic statements such as strict laws or near-laws (in the sense just defined) play. Rather, a "*ceteris paribus* law" is an element of a "work in progress," an embryonic theory on its way to being developed to the point where it makes definite claims about the world. It has been found that in a vaguely defined set of circumstances, a given generalization has appeared to be mostly right or mostly reliable, and there is a hunch that somewhere in the neighborhood is a genuine, well-defined generalization, for which the search is on. But nothing more precise than this can be said, yet. To revive a now-unfashionable notion, "*ceteris paribus* laws" belong to the context of discovery rather than the context of justification. And while we do not adhere to the old logical empiricist dictum that philosophy is to restrict itself to considering the context of justification, we do submit that the philosophical analysis that is called for in the case of a "work-in-progress" theory is probably quite different from that called for in the case of a "finished" theory that already makes definite claims about the world and so is a candidate for empirical confirmation or disconfirmation. In particular, whereas we see the need for a philosophical analysis of the truth conditions of strict law-statements and near-law-statements, and a philosophical account about how such statements can be confirmed, we do not see any such needs in the case of "*ceteris paribus* laws." This is because we maintain that such "laws" are only vague statements that partly constitute embryonic theories; they are not put forward as true (except perhaps in the vague and Pickwickian sense that something in their neighborhood is probably true) and there appears to be little interest in an account of what the world would have to be like for them to be true, or to be "*ceteris paribus*" true (whatever that might mean). While there are probably many interesting things to say about the relation between "*ceteris paribus* laws" and the evidence available in the sciences that sport them, we claim that this relation must be quite different from that studied in confirmation theory, where hypotheses that make definite claims about the world are tested against empirical data. Likewise, whatever explanatory power such "laws" have is probably quite different from that of precisely formulated law-statements.[43]

If laws are needed for some purpose, then we maintain that only laws will do, and if "*ceteris paribus* laws" are the only things on offer, then what is needed is better science, and no amount of logical analysis on the part of philosophers will render the "*ceteris paribus* laws" capable of doing the job of laws. Perhaps there are purposes for which laws are not needed, and "*ceteris paribus* laws" will serve,[44] but since we maintain that "*ceteris paribus* laws" are inherently vague and without definite truth conditions, we think it follows that in any such situations, a true account of the world is not needed. If there are such situations, then perhaps they are situations where the theories needed are best given an instrumentalist construal. But we suspect that the main interest of "*ceteris paribus laws*" is that they are (hopefully) stations on the way to a better theory with strict generalizations (or at any rate, statements with precise contents). The challenge for philosophers of science here is to understand this process; it is tempting but dangerous to mistake the way-station for the destination, and to attempt to analyze "*ceteris paribus* laws" in a way that minimizes or obscures their differences from strict laws.

12. The Illusion of the Importance of *Ceteris Paribus* Laws for the Special Sciences

We fear that our remarks in the previous section may seem to invite charges of "physics chauvinism." To say that if laws are needed, then "*ceteris paribus* laws" will not do, and if only "*ceteris paribus* laws" are in hand then we need better science, looks at first glance to be a negative judgment about the special sciences as compared with fundamental physics. In this section we shall try to dispel this appearance.

Toward this end, let us return to Kincaid's (1996) discussion of nine methods of confirming *ceteris paribus* laws. Again, we think that these methods embody much good scientific sense; but we wish to consider the question of what, exactly, is the content of the claims that are tested by such methods. One answer is that these methods test claims about actual correlations among variables across various populations. For example:

> (5) In population H, P is positively statistically correlated with S across all sub-populations that are homogeneous with respect to the variables $V_1, \ldots, V_n$.

(5) does not suffer from the vagueness of "*ceteris paribus* laws" (so long as the variables *P, S* and $V_1 \ldots V_n$, are defined precisely enough). It asserts a certain precisely defined statistical relation among well-defined variables. Most of the methods described by Kincaid seem to be good methods for testing claims like (5). Recall that one of Kincaid's methods involves looking to cases in which "the *ceteris paribus* conditions are satisfied."[45] As we noted above, this requires that we be able to *tell* when these are satisfied. This suggests that the *ceteris paribus* clause is not a vague and open-ended "escape clause." The clauses Kincaid has in mind might well be taken to be such clauses as, "so long as variable V_i has been controlled for." In this case, the method Kincaid describes is essentially that of controlling for relevant variables, which is certainly a reasonable practice if one is concerned to test a claim of the form (5). Similar remarks apply to some of Kincaid's other methods.

Kincaid argues that his methods are exemplified in some good work in the social sciences. In particular, he provides an illuminating discussion of Paige's (1975) work on agrarian political activity.[46] If we let H stand for the population of humans between 1948 and 1970, *P* stand for degree of political activity among cultivators, *S* stand for economic relations among non-cultivators and cultivators in agrarian societies, and the V_i's stand for such factors as the proximity of progressive urban political parties, then (5) closely approximates the results of Paige's statistical analysis, which was carried out using many of Kincaid's methods. This at least suggests the idea that what really gets tested by these methods tends to be propositions about statistical distributions conditional on controlled variables, such as (5). Of course, much research in the special sciences produces results of just this form.

Kincaid argues forcefully that the results of careful applications of the methods he lists, such as Paige's work, can have great explanatory value. We agree, and in particular we think that this can be seen to be the case if we suppose that the results of such research typically take the form of (5). According to Kincaid, Paige's work allows us to infer much about the (partial) causes of particular political events, and that such information has explanatory import. We don't wish to get into the debates about the notion of causality here, but Kincaid's point seems to be very plausible. Moreover, as many philosophers have argued, there seems to be no compelling reason to suppose that in order to shed light on the causes of individual events, it is necessary to cite any general laws.[47] So (5) could be useful for providing causal explanations even if it doesn't state or imply a law.

Furthermore, even for someone wary of claims about causality, there is still some reason to think that propositions like (5) might have explanatory virtues. For example, (5) seems to provide exactly the kind of information that would be crucial to constructing a statistical explanation for the degree of political activity in a particular agrarian community, on Salmon's S-R model[48] or some similar alternative. Again, this way of understanding the explanatory import of (5) needn't involve any claim about laws, since the S-R model does not specifically require laws in the explanans.

So it seems quite plausible that a claim like (5), which can be tested using Kincaid's methods, can have explanatory import, even if there is no particular nomological claim licensed by (5). But another way of understanding the explanatory import of (5) has it that (5) permits us to infer the existence of a law of nature, and that this law has explanatory value precisely because it can figure in D-N or covering-law explanations. This line of thought goes hand-in-hand with a second way of answering the question of exactly what claim is tested by Kincaid's nine methods. The second answer, of course, is that what is tested is a putative law. Apparently, the law in question is not a strict law—this would involve the claim that in all other (nomologically possible) populations of the same kind, the same variables would be positively correlated, when the same other variables are controlled for. As is a familiar point by now, the systems studied by the special sciences tend to be too dependent upon contingency and circumstance for there to be such strict laws. So the putative law must be a putative *ceteris paribus* law. The idea here is essentially that if methods like those described by Kincaid can establish a statistical claim like (5), then from this we may infer:

(6) *Ceteris paribus, P* and *S* covary.

The "*ceteris paribus*" clause here covers significant variations in the variables $V_1 \ldots V_n$, but also other possible interfering factors, not all of which may be explicitly formulated.

On this line of thought, we are forced to face the first horn of the dilemma of Pietroski and Rey (see Section 5). That is to say, if we suppose that this is in fact the way to understand what is tested by Kincaid's methods, and if we suppose that the explanatory power of the results of these methods is to be understood in terms of the ability of a law of nature to figure in covering-law explanations, and if we insist that the legitimacy and explanatory import of the special sciences must be respected, then it follows that we must understand *ceteris paribus* laws as genuine laws. But as we have argued, the prospects here are pretty bleak. It does not follow that Kincaid's methods are not legitimate and scientific, nor does it follow that the results of these methods lack explanatory value. All that follows is that *one* way of understanding these methods and the value of their results leads to a dead end.

We have considered two ways of understanding the role played by Kincaid's methods in the practice of science. On the first way, these methods are useful ways of confirming claims like (5), and (5) has explanatory value, either because it sheds light on particular causal histories, or because it provides statistical data useful for providing statistical explanations, or both. On the second way, Kincaid's methods can be used to confirm (5), but this is not the inferential stopping point: (5) is used as a premise for an ampliative inference to (6), and the conclusion of this inference can be used to give covering-law explanations of, among other things, (5) itself. We object to this second line of thought, not because we have epistemological doubts about the rationality of the supposed ampliative inference, but because we think that the supposed conclusion of this supposed inference is empty.[49] Thus, we favor the first way.[50]

Let us return to our slogan, "If laws are needed, then only laws will do, and *'ceteris paribus* laws' will not." The point of this slogan is not that the special sciences cannot be scientifically legitimate. Rather, the point is that when only "*ceteris paribus* laws" are on offer, then whatever scientific purposes are being fulfilled (and, as Kincaid's discussion of Paige illustrates, important scientific purposes can be fulfilled in such cases) do not require laws. Hence, there is no need to try to rescue the special sciences by finding a way to minimize the differences between "*ceteris paribus* laws" and laws. To do so is to try to stuff all good science into the pigeon hole modeled on fundamental physics, which, we have argued, does articulate strict laws of nature. This can only obscure what is important about the special sciences, as well as what is important about fundamental physics.

We agree with Kincaid that his nine methods are reasonable ones for scientists to employ. We find it plausible that the results of such methods may have explanatory value. So we think that we agree with the main point that Kincaid argues for in the passages we have considered—namely, that social science can be legitimate science. What we object to is one gloss that Kincaid sometimes puts on his position: that the social sciences articulate *ceteris paribus* laws, and that such laws play a legitimate role in scientific practice.[51] (We should note that this gloss doesn't seem to be essential to the main thrust of Kincaid's argument, and at some places he distances himself from it.[52]) The gloss strikes us as unfortunate, because it gives aid and comfort to philosophical projects for explicating "*ceteris paribus* laws" in a way that minimizes their glaring difference from strict laws of nature. As we have explained, we find such projects hopeless, misguided, and irrelevant to understanding what really goes on in science.

13. Conclusion

We have argued that it isn't *ceteris paribus* all the way down—*ceteris paribus* stops at the level of fundamental physics. Furthermore, given Hempel's insight, if all regularities in the world supervene on the regularities that can be studied by physics, then there can be no strict laws of a distinctively biological, psychological, or economic kind; that is, there can be no strict laws formulated purely within the vocabulary of the special science in question. The supervenience claim may be challenged, but most writers on the topic seem to be in agreement with the conclusion that there are no strict laws of the special sciences. It follows that if there are laws of the special sciences—and many commentators assume that there must be—then there must be *ceteris paribus* laws. But there is no persuasive analysis of the truth conditions of such laws; nor is there any persuasive account of how they are saved from vacuity; and, most distressing of all, there is no persuasive account of how they meld with standard scientific methodology, how, for example, they can be confirmed or disconfirmed. In sum, a royal mess.

Our rhetorical strategy for finding a way out of this mess was to start out to answer the question "What is a *ceteris paribus* law?" to provide an answer by giving truth conditions, then to see what is wrong with the answer, and, finally, to draw the moral that the question is not a good one. There is a well-defined sense in which a generalization can fail to be strictly true and yet be a "near-law"; we have shown how the "laws" of classical phenomenological thermodynamics answer to such a sense. Where a micro-reduction is possible, it is hopeful that such a clear sense of "near law" can be satisfied. But what is crucially important is that there be precise senses that can be given to "approximately true" and "most of the intended applications." In most cases where *ceteris paribus* clauses are actually used, this is not the case, and in our example of a situation where this is the case, *ceteris paribus* clauses are not used. Our positive proposal is that when the requisite precise senses cannot be defined, "*ceteris paribus* laws" are the vague claims that they appear to be, and that their widespread use can be explained by the fact that they are elements of "work-in-progress theories." When they are put forward by a science, this is an indication that science is still in the process of elaborating a theory that makes definite claims about the world; philosophers should let the scientists get on with their work and try to understand this process, rather than attempting to analyze "*ceteris paribus* laws" in a way that hides their shortcomings and obscures the road that lies ahead for science.

As we hope we have made clear, we think there is much more to the special sciences than just articulating such "work-in-progress" theories. Nonetheless, articulating such embryonic theories might be a real feature of scientific practice, in both the special sciences and physics, and we suspect that it is. So "*ceteris paribus* laws" might have a place in an adequate understanding of science, although we think it must be quite different from that typically ascribed to them. For they might be elements of embryonic theories. As such, they are not yet ready to be confirmed or disconfirmed, and it is not clear that they can have real explanatory import in their current stage of development. Thus, they do not stand in need of the same kind of explication as do the propositions of fully formed theories, such as laws, conditional probability statements, and the like. And it would be a mistake to try to analyze them in such a way as to obscure or minimize the ways in which they differ from the latter.

Much work on the topic of provisos and *ceteris paribus* laws has been motivated by a concern to defend the special sciences. The concern often derives from the following line of reasoning: "These sciences do not state strict laws, so they must state *ceteris paribus* laws; the scientific status of these sciences is not to be impugned, so we must find a way of showing that *ceteris paribus* laws are not really that different from the laws of fundamental physics." We remain "physics chauvinists" in the limited sense that we *do* think there is a crucial difference here. It is not "*ceteris paribus* all the way down"—*ceteris paribus* stops at the level of fundamental physics. But we are *not* physics chauvinists in a more important sense, for we deny that the mark of a good science is its similarity to fundamental physics. The concept of a law of nature seems to us to be an important one for understanding what physics is up to, but it is a misguided egalitarianism that insists that what goes for physics goes for all the sciences. The special sciences need not be in the business of stating laws of nature at all, and this blocks the inference from the legitimacy of these sciences to the legitimacy of *ceteris paribus* laws. For us, it is ironic that an effort to justify the special sciences takes the form of trying to force them into a straitjacket modeled on physics. We think this effort should be resisted, since it damages both our understanding of the special sciences and our understanding of the concept of a law of nature.

Notes

1. This claim is made by Cartwright (1983, 1989, 1995, 1997), Fodor (1991), Giere (1988), Hausman (1992), Kincaid (1990, 1996), and many of the other works cited below—but *not,* we will argue (against prevailing opinion), by Hempel (1988).

2. The list of references given below is by no means complete, but it does provide the reader with a representative sample of the recent work on this problem. For a history of the origins and use of *"ceteris paribus,"* see Persky (1990).

3. Lange (1993a, 233). Later we will question the efficacy of this example.

4. Here Kincaid is using "*ceteris paribus* clause" to mean the same thing that Lange means by "proviso." The two are often used interchangeably in the literature, and generally we will follow this practice. However, as will be explained below, Hempel uses proviso in a different sense.

5. Lakatos (1970) also held that theories of physics contain *ceteris paribus* clauses. But

his sense of *ceteris paribus* seems closer to Hempel's sense of proviso discussed below in Section 3 than to the standard sense of *ceteris paribus.*

6. This holism results from the notion that the HD method is all there is to inductive reasoning, a very dubious notion indeed. However, we do not wish to inveigh against holism here.

7. See especially chapter 4 ("The Language of Theories") of Sellars (1963, 1991).

8. Thus, Lange (1993a, 240): "the number of provisos is 'indefinitely large,' which makes it impossible to offer them all as premises."

9. Lange's own solution to the problem of provisos shares this feature of Hempel's account. As will become clear below, Lange holds that the content of a scientific theory typically cannot be explicated in terms intelligible prior to the introduction of the theory. So it isn't that Lange completely misses Hempel's point; it's just that he takes the main upshot of Hempel's argument to be something other than what we take it to be, and what Hempel says it is.

10. In particular, see Hempel (1988, 26): "Note that a proviso as here understood is not a clause that can be attached to a theory as a whole and vouchsafe its deductive potency. . . . Rather, a proviso has to be conceived as a clause that pertains to some particular application of a given theory."

11. To be fair to Lange, it should be noted that he acknowledges that Hempel does not explicitly present the dilemma discussed above. But he claims that "this dilemma certainly stands behind [Hempel's] discussion" (1993a, 238). He argues that "[i]t must be because he believes that a 'law-statement' without provisos would be false, that Hempel defines provisos as 'essential'" (ibid.). But as we have pointed out, Hempel does not think that without a proviso, a law would be false. In the Newtonian celestial mechanics example, Hempel takes the laws to be true as they stand. He argues that the proviso that there are no non-negligible forces other than mutual gravitational attraction is essential because it is necessary for *this particular application of the laws,* not because it is essential to the truth of the laws. It might be replied that Hempel must take the laws in this example to include such clauses as "so long as no unaccounted-for forces are acting, but the work done by this clause is already done by the reference, in Newton's second law of motion, to the *total* impressed force (see below).

12. This is not to say that the semantic view of theories does not have advantages over the statement view of theories, but our opinion (for which we will not argue here) is that the virtues of the semantic view have been greatly exaggerated.

13. By "phenomenological physics" we mean those branches of physics that aim to state correlations among more or less observable macroscopic phenomena: hence, Lange's example of the law of thermal expansion belongs to phenomenological physics. By the "special sciences" we mean to include all sciences other than physics; but we have in mind particularly biology, psychology, and economics.

14. Toward this end we have to take on what would be Cartwright's (1983) objection to our analysis of Hempel's example, and in particular, to our claim that Newton's law of gravitation stands or falls without proviso or *ceteris paribus* qualification. There are two ways to construe this law. One is to take it as asserting that "If there are no non-gravitational forces acting, then any two massive bodies exert a force on one another directly

proportional to the product of their masses and inversely proportional to the square of the distance between them." The other is to take it as asserting that "(Regardless of what other forces may be acting) any two massive bodies exert a gravitational force on one another that is directly proportional to the product of their masses and inversely proportional to the square of the distance between them." Cartwright grants that the first reading states (what was taken to be) a true law; but she notes, quite correctly, that this law is irrelevant to real world situations where typically other forces are present. The second reading (which we favor) produces a more useful law statement but one that according to Cartwright lacks facticity because component forces are unreal. We first observe that even if correct, Cartwright's view is not very damaging to our thesis because there are plenty of other examples that are not subject to her peculiar form of anti-realism about component forces (e.g., Einstein's general theory of relativity which treats gravity in terms of spacetime curvature rather than in terms of force). Furthermore, we do not understand how anything short of a blanket anti-realism can motivate the notion that the gravitational force component of a total impressed force is unreal. To be sure, it is not implausible to say that an arbitrary decomposition of the total resultant force may yield components to which we may not want to assign any direct ontological significance. But modern physical theory from Newton onward gives two reasons to take certain component forces as having real ontological significance: first, the theory gives an account of how the component force arises from the distribution of sources (masses for the gravitational force, charges for the electrical force, etc.); and it promotes a form of explanation in which the total resultant force is obtained as a vector sum of the component forces that are due to sources.

15. But since these generalizations are, by hypothesis, true, the problems of *ceteris paribus* and provisos would not arise. As a matter of fact, however, there seem to be very few strictly true generalizations which can be stated purely in the vocabulary of a special science and which are lawlike and play in that special science the sorts of roles that would make it plausible to dub them laws of this special science.

16. Admittedly there is much more to Beatty's thesis, which we cannot do justice to here. In particular, he shows that there are various senses in which the "laws of biology" are deeply contingent.

17. Though not unprecedented—see Musgrave (1980) and the references therein.

18. This isn't Lange's entire case—he provides other arguments in his (1993b) and (1995). Here we cannot do justice to all of Lange's views and arguments concerning laws. Our concern here is not to criticize his positive account of laws, but only to object to his treatment of the problem of provisos in his (1993a).

19. See Musgrave (1980) for arguments against the view of laws as rules.

20. Here and subsequently, "cp: P" will be used to express the proposition that P holds under a *ceteris paribus* clause.

21. Actually, Pietroski and Rey say that they are giving a condition for the *nonvacuity* of a *ceteris paribus* law. But it is clear that their sufficient condition for nonvacuity will not be satisfied by putative *ceteris paribus* laws that are nonvacuous but *false*. Hence, we construe their proposal as a proposed sufficient condition for such a law to be both true and nonvacuous.

22. Pietroski and Rey are non-committal about this third clause; its purpose is to rule out laws that are empty in the sense of having neither any instances nor any exceptions. They allow for the possibility that this clause should be dropped. We have corrected what we take to be a typo in clause (ii). In the published version, this clause ends with the phrase "explains [—*Gz*]." But this "z" is a variable unbound by any quantifier. So we read "y" for "z."

23. Cartwright draws a distinction between capacities and tendencies, but this distinction will not matter for our discussion; see her (1989, 226).

24. Cartwright (1995, 293): "Economics and physics equally employ *ceteris paribus* laws, and that is a matter of the systems they study, not a deficiency in what they say about them."

25. See Cartwright (1989), 158–70.

26. Cartwright (1989, 162–63). For Sellars's argument, see his (1963, 95–97 and 118–23).

27. Sellars uses his argument to establish that theoretical laws typically do not have any contingent consequences that can be stated purely within the "observation framework." It does not follow from this that there are no strict lawlike regularities governing observable events; indeed, Sellars is concerned to argue that one important function of theories positing unobservables is that they allow us to formulate strictly true lawlike generalizations where before we could not (see the references in the last note). It seems to us that Cartwright's parallel argument may well establish that attributions of capacities do not imply any strict regularities that can be stated in a vocabulary without the resources to refer to capacities; Cartwright concludes, however, that there are no strict lawlike regularities in nature at all, not even ones that can be only be stated in a richer vocabulary that mentions capacities. This is the conclusion we are taking issue with.

28. Thus, she writes, in her (1989, 206–07): "The abstract law [i.e., the fundamental law ascribing capacities] is one which subtracts all but the features of interest. To get back to the concrete laws that constitute its phenomenal content [i.e., whatever observable regularities the abstract law gives rise to in concrete situations], the omitted factors must be added in again. But where do these omitted factors come from? . . . given a theory, the factors come *from a list.* But the list provided by a given theory, or even by all of our theories put together, will never go far enough. There will always be further factors to consider which are peculiar to the individual case."

29. The sketch just provided is an informal account of what Cartwright describes as the process of concretization in her (1989, 202–6). The problem of the factors peculiar to the given context is what she calls the "problem of material abstraction" on p. 207; it is introduced by the passage quoted in the preceding note.

30. Cartwright (1983, 19).

31. Silverberg's analysis is directed at cp claims of the form cp: $(A \rightarrow B)$ rather than cp: $((x)(Ax \rightarrow Bx))$ although the latter is more relevant to cp laws. Presumably he would say that the latter is true iff $(x)(Ax \rightarrow Bx)$ is true in all possible worlds that are appropriately ideal and are otherwise most similar to the actual world.

32. This analysis has been influential in the philosophy of economics: see, for example, Rosenberg (1992).

33. This needs to be qualified: As it stands, Hausman's proposal is trivial, since there always exists a condition *C* that will do the job, namely *G* itself. But *G* would not count as completer in Fodor's sense, because a completer must not by itself be a sufficient condition for the consequent of the law. So by calling *C* a completer, we are implicitly building in a correction that Hausman's analysis needs anyway.

34. See Hausman (1991, 139–42).

35. For criticisms of Lewis's account, see van Fraassen (1989) and Carroll (1994). We are optimistic that these criticisms can be met, but one of us (J. R.) thinks that Lewis's account fails for other reasons.

36. At any rate, this treatment is more in keeping with Hempel's (1988) discussion (see especially the discussion of Newtonian celestial mechanics on p. 23) than is Lange's discussion.

37. That this is not an idle point, see Hausman (1991, 135 n. 13), who claims the opposite.

38. For instance, a dynamical theory will be deemed approximately true if the trajectories predicted by the theory track actual trajectories sufficiently closely. We are in complete agreement with Peter Smith (1998) that for dynamical theories, Popper's notion of verisimilitude, which gauges nearness to the truth in terms of the amount of exact truths that are captured, is badly off the mark. A dynamical theory can be approximately true in the above sense even if *all* of its assertions about trajectories are strictly false, and it can be nearer the truth than some other theory which makes many more strictly true statements about trajectories but which fails to track the actual trajectories as closely.

39. For example, suppose for sake of argument that "*ceteris paribus,* if demand increases, then the price increases" is a cp law of economics. Now suppose that the boundaries of the sciences shift so that economics makes use of the vocabulary and experimental techniques of micro-physics. It seems plausible that this "law" will no longer be a law of economics since economics will now be able to study many kinds of micro-physical states that might well defeat the "law."

40. There are unresolved technical and conceptual problems in the reduction of thermodynamics to statistical mechanics (see Sklar [1993]), but these problems do not affect the present discussion.

41. See Earman and Norton (1998) for an account of this matter.

42. Carrier (1998) calls the second law of thermodynamics a *ceteris paribus* law because it is "afflicted with exceptions" (221). But when this generalization was formulated in the nineteenth century, it was believed to hold without exception. And when exceptions were discovered in the twentieth century, scientists typically reacted not by redubbing it a *ceteris paribus* law but by adding (implicit) scare quotes to the honorific "law" and noting that it is only an approximately true generalization whose limited reliability is to be explained by statistical mechanics.

43. A helpful analogy is provided by Faraday's striking and imaginative statements about lines of force. Such statements pointed the way to the laws of electrodynamics, but they were not such laws themselves, and it may be appropriate to think of them as part of a "work-in-progress" theory. "What are the truth conditions of these statements? What is their precise content? How can we make sense of the way they are confirmed, and the

role they play in making predictions and giving explanations?" What would be the point of asking questions like these? But surely there are more interesting questions in the neighborhood, e.g., "What role did Faraday's claims about lines of force play in the development of Maxwell's theory of electromagnetic fields?"

44. Another possible case, of course, is one in which some legitimate scientific purpose requires neither laws nor "*ceteris paribus* laws." See Section 12.

45. Kincaid (1996, 67).

46. Kincaid (1996, pp. 70–80).

47. Anscombe (1971); Cartwright (1983).

48. Salmon (1970).

49. What we think is empty is not the notion of a law as such, but the notion of a *ceteris paribus* law.

50. Of course, someone could say that the conclusion of the kind of investigation Kincaid describes is a statement of the form (6), but that such statements are just convenient shorthand for statements of the form (5). If this is done, then nothing is amiss, but we would say that "*ceteris paribus*" is being used in the *lazy* sense described in Section 11, rather than in the philosophically interesting sense that has been the focus of most of the literature on this topic.

51. Kincaid (1996, 63ff).

52. For example, on p. 97 of his (1996), Kincaid writes: "[I]n the end the real explanatory work results from picking out the particular causes at work. Generalizations help in that process and result from it, but they are really derivative of the specific causal facts. Thus though I used causal laws as a wedge into this chapter, the key factor in explanation, I would suggest, is not the laws but the causes. . . . Laws . . . are likely to pick out only very partial causes. They will be confirmed and will explain only to the extent that we are sure they apply—and that is done best by filling in their *ceteris paribus* clauses, frequently on a case-by-case basis." Thus, Kincaid shares our skepticism that the explanatory fruits of his nine methods are covering-law explanations. He also seems to share our skepticism that "*ceteris paribus* laws" (which are clearly the only kind of laws he is concerned with here) have explanatory import and are capable of being confirmed as long as their *ceteris paribus* clauses are left vague.

References

Anscombe, G. E. M.: 1971, *Causality and Determinism,* Cambridge University Press, Cambridge.

Beatty, J.: 1995, "The Evolutionary Contingency Thesis," in J. Lennox et al. (eds.), *Concepts, Theories and Rationality in the Biological Sciences,* University of Pittsburgh Press, Pittsburgh, PA, 45–81.

Carrier, M.: 1998, "In Defense of Psychological Laws," *International Studies in the Philosophy of Science* 12, 217–32.

Carroll, J.: 1994, *Laws of Nature,* Cambridge University Press, Cambridge.

Cartwright, N.: 1983, *How the Laws of Physics Lie,* Oxford University Press, Oxford.

Cartwright, N.: 1989. *Nature's Capacities and Their Measurement,* Oxford University Press, Oxford.

Cartwright, N.: 1995, "*Ceteris Paribus* Laws and Socio-Economic Machines," *Monist* 78, 276–97.

Cartwright, N.: 1997, "Where Do the Laws of Nature Come From?" *Dialectica* 51, 65–78.

Earman, J. and Norton, J. D.: 1999, "Exorcist XIV: The Wrath of Maxwell's Demon," *Studies in History and Philosophy of Modem Physics,* in press [vol. 30, 1-40—*JWC*].

Fodor, J. A.: 1991, "You Can Fool Some of the People All the Time, Everything Else Being Equal: Hedged Laws and Psychological Explanations," *Mind* 100, 19–34.

Giere, R.: 1988, "Laws, Theories, and Generalizations," in A. Grunbaum and W. Salmon (eds.), *The Limits of Deductivism,* University of California Press, Berkeley, CA, 37–46.

Hausman, D. M.: 1992, *The Inexact and Separate Science of Economics,* Cambridge University Press, Cambridge.

Hutchison, T. W.: 1938, *The Significance and Basic Postulates of Economic Theory,* Macmillan, London. Reprinted 1965. New York: A. M. Kelly.

Hempel, C. G.: 1988. "Provisos: A Problem Concerning the Inferential Function of Scientific Laws," in A. Grunbaum and W. Salmon (eds.), *The Limits of Deductivism,* University of California Press, Berkeley, CA, 19–36.

Kincaid, H.: 1990, "Defending Laws in the Social Sciences," *Philosophy of the Social Sciences* 20, 56–83.

Kincaid, H.: 1996, *Philosophical Foundations of the Social Sciences,* Cambridge University Press, Cambridge.

Lakatos, I.: 1970, "Falsification and the Methodology of Scientific Research Programs," in I. Lakatos and A. Musgrave (eds.), *Criticism and the Growth of Knowledge,* Cambridge University Press, Cambridge, 91–195.

Lange, M.: 1993a, "Natural Laws and the Problem of Provisos," *Erkenntnis* 38, 233–48.

Lange, M.: 1993b, "Lawlikeness," *Nous* 27, 1–21.

Lange, M.: 1995, "Are There Natural Laws Concerning Particular Biological Species?" *Journal of Philosophy* 92, 430–51.

Lewis, D.: 1973, *Counterfactuals,* Harvard University Press, MA.

Musgrave, A.: 1980, "Wittgensteinian Instrumentalism," *Theoria* 45 (6), 65–105.

Mott, R: 1992, "Fodor and Ceteris Paribus Laws," *Mind* 101, 33–346.

Paige, J.: 1975, *Agrarian Revolutions,* Free Press, New York.

Persky, J.: 1990, "Ceteris Paribus," *Journal of Economic Perspectives* 4, 187–93.

Pietroski, R., and Rey, G.: 1995, "When Other Things Aren't Equal: Saving *Ceteris Paribus* Laws from Vacuity," *British Journal for the Philosophy of Science* 46, 81–110.

Rosenberg. A.: 1992, *Economics: Mathematical Politics or Science of Diminishing Returns?* University of Chicago Press, Chicago.

Salmon, W.: 1970, "Statistical Explanation," reprinted in W. Salmon (ed.), *Statistical Explanation and Statistical Relevance,* University of Pittsburgh Press, Pittsburgh, 1971.

Schiffer, S.: 1991, "Ceteris Paribus Laws," *Mind* 100, 1–17.

Sellars, W.: 1963, *Science, Perception and Reality,* Humanities Press, New York. Reprinted 1991, Ridgeview Press, Atasacadero, CA.

Silverberg, A.: 1996, "Psychological Laws and Non-Monotonic Logic," *Erkenntnis* 44, 199–224.

Sklar, L.: 1993, *Physics and Chance,* Cambridge University Press, Cambridge.

Smith, R.: 1998, *Explaining Chaos,* Cambridge University Press, Cambridge.

Van Fraassen, B. C.: 1980, *The Scientific Image,* Clarendon Press, Oxford.

Van Fraassen, B. C.: 1989, *Laws and Symmetries,* Clarendon Press, Oxford.

12

The Non-Governing Conception of Laws of Nature

HELEN BEEBEE

I. Introduction

There are two main camps in the debate about the metaphysics of laws of nature. In one corner, there is the anti-Humean view of David Armstrong: laws are relations of necessity between universals.[1] And in the other corner there is the Ramsey-Lewis view: laws are those generalizations which figure in the most economical true axiomatization of all the particular matters of fact that obtain.[2] The Ramsey-Lewis view counts as a Humean view because it does not postulate any necessary connections. The debate between the rival camps can be read as a debate about whether or not supervenience holds for laws of nature: whether or not nomic facts supervene on non-nomic facts or, to put it in more Lewis-esque terms, whether or not laws supervene on the overall distribution of particular matters of fact.[3]

It's worth pointing out that in principle the upholder of supervenience with respect to laws has two options: she can view the supervenience of laws on particular matters of fact as either contingent or necessary. That is, she can hold that laws of nature supervene on particular matters of fact only at our world and worlds reasonably similar to ours, or, more strongly, that it's true at *all* possible worlds that laws supervene on particular matters of fact. Indeed, it's quite easy to imagine the form which a contingent supervenience thesis might take. Taking our cue from functionalism in the philosophy of mind, we might try to identify a "folk theory" of lawhood, analogous to folk theories of, say, pain and belief. The literature on laws of nature points to a good deal of agreement about what a "folk theory" of laws of nature might look like (laws are those features of the world which make true universal, counterfactual-supporting generalizations; laws are explanatory; and so on). And we could then use the folk theory to pick out

From *Philosophy and Phenomenological Research* 61 (2000): 571–94.

laws of nature non-rigidly, so that at our world laws of nature are those generalizations which figure in the best axiomatic system, whereas at other worlds laws are relations of necessity between universals, or the rules God uses for deciding how the universe is going to evolve, or whatever.

As far as I know, nobody has tried to articulate or defend the contingent supervenience view (though it might turn out to be an attractive position for those Humeans who are more concerned than I am about the thought experiments discussed in sections V and VI below). I shall therefore ignore the contingent supervenience option for the remainder of this paper, and identify Humeanism with the stronger, reductionist thesis that laws supervene on particular matters of fact at all possible worlds. With Humeanism thus construed, the debate about laws of nature is a classic realism-versus-reductionism dispute, with the anti-Humean realists wanting to ground the distinction between laws and accidentally true generalizations in some metaphysically substantive feature of the world—something irreducibly nomic—and the Humean reductionists wanting to preserve the law-accident distinction without recourse to any suspiciously anti-Humean ontology.

A fairly standard way of criticizing the Humean view has been the use of thought experiments; this is a method employed by, for instance, Bas van Fraassen, Michael Tooley, and John Carroll.[4] We are asked to consult our intuitions about some remote possible world, and then shown how those intuitions conflict with the verdict of Humeanism about laws (or more specifically with the verdict of the Ramsey-Lewis view). My purpose in this paper is to show that two such thought experiments do not succeed in finishing off the Humean conception of laws, because they presuppose a conception of laws which Humeans do not share: a conception according to which the laws *govern* what goes on in the universe.

My strategy is going to be as follows. First of all (in section II) I'll give a brief characterization both of the Ramsey-Lewis view and of Armstrong's view, and show how they both do justice to some of our more obvious common-sense intuitions about the nature of laws. In section III, I discuss what I see as the fundamental difference between the two views, which is a difference over whether or not laws *govern*. I bring out the distinction between the "governing" conception favored by anti-Humeans and the "descriptive" conception favored by Humeans by showing how the distinction informs debates about free will and determinism. The intuition that laws govern is, I think, deeply felt—at least implicitly—by a lot of philosophers, and probably by a lot of the folk too. But it is an intuition that the Ramsey-

Lewis view—and Humeanism about laws in general—refuses to endorse. And it's no accident that it refuses to do so: the intuition that laws govern is precisely the intuition which leads one to postulate the necessary connections—as ontological grounds of the governing nature of laws—which the Humean refuses to allow into her ontology.

The prevalence of the view that laws play a governing role suggests a quick refutation of Humeanism: if it is a *conceptual* truth that laws govern, then Humeanism, which accords laws of nature no such status, must be false on conceptual grounds. In section IV, I argue that it is *not* a conceptual truth that laws of nature govern, since the only motivation for the claim that it *is* a conceptual truth is a false analogy with other kinds of law.

The point of sections III and IV is to show that the Ramsey-Lewis view is, at least *prima facie,* a coherent, well-motivated view—and one which cannot be dismissed simply on the grounds that it fails to accord laws of nature a governing role. In sections V and VI, I look at two thought experiments that are supposed to be counter-examples not just to the Ramsey-Lewis view but to Humeanism about laws in general. I argue that the alleged "common-sense" intuitions which the counter-examples appeal to—the intuitions that are supposed to refute Humeanism—are explicitly anti-Humean intuitions, and therefore just the sort of intuition to which the dedicated Humean ought not to be doing justice. In particular, the primary anti-Humean intuition appealed to in the thought experiments is the intuition that laws of nature govern. But if the argument of section IV is right, this is an optional intuition; it is one which the Humean need not, and indeed should not, share. Thus Humeanism about laws—and the Ramsey-Lewis view in particular—is not refuted by the thought experiments.

I'm going to assume determinism in what follows. I dare say this is a false assumption, but it is one that is warranted by the purposes of this paper, partly because the counter-examples discussed below assume determinism, and partly because matters are very much more complicated without the assumption. However, while the details differ once the assumption of determinism is dropped, the fundamental dispute about whether or not laws govern remains. Under indeterminism, the anti-Humean will say that laws govern chances of events rather than events themselves, or that probabilistic laws govern only in those cases where the consequent of the law is realized, or some such. On the other hand, the Humean will say that laws are merely true general *descriptions* of the chances of events (where "chance" is given a suitably Humean reading), or that laws describe not constant co-occurrences but merely more or less frequent co-occurrences.

II. Two Theories of Lawhood

The Ramsey-Lewis View

In a short note written in 1928, Ramsey defined lawhood like this: "if we knew everything, we should still want to systematize our knowledge as a deductive system, and the general axioms in that system would be the fundamental laws of nature" (1978, p. 131). In *Counterfactuals,* Lewis recasts the definition as follows: "a contingent generalization is a *law of nature* if and only if it appears as a theorem (or axiom) in each of the true deductive systems that achieves a best combination of simplicity and strength" (1973, p. 73).

So the idea is something like this. Suppose God wanted us to learn all the facts there are to be learned. (The Ramsey-Lewis view is not an epistemological thesis but I'm putting it this way for the sake of the story.) He decides to give us a book—God's Big Book of Facts—so that we might come to learn its contents and thereby learn every particular matter of fact there is. As a first draft, God just lists all the particular matters of fact there are. But the first draft turns out to be an impossibly long and unwieldy manuscript, and very hard to make any sense of—it's just a long list of everything that's ever happened and will ever happen. We couldn't even come close to learning a big list of independent facts like that. Luckily, however (or so we hope), God has a way of making the list rather more comprehensible to our feeble, finite minds: he can axiomatize the list. That is, he can write down some universal generalizations with the help of which we can derive some elements of the list from others. This will have the benefit of making God's Big Book of Facts a good deal shorter and also a good deal easier to get our rather limited brains around.

For instance, suppose all the facts in God's Big Book satisfy $f = ma$. Then God can write down $f = ma$ at the beginning of the book, under the heading "Axioms," and cut down his hopelessly long list of particular matters of fact: whenever he sees facts about an object's mass and acceleration, say, he can cross out the extra fact about its force, since this fact follows from the others together with the axiom $f = ma$. And so on. God, in his benevolence, wants the list of particular matters of fact to be as short as possible—that is, he wants the axioms to be as strong as possible; but he also wants the list of *axioms* to be as short as possible—he wants the deductive system (the axioms and theorems) to be as simple as possible.[5] The virtues of strength and simplicity conflict with each other to some extent; God's job is to strike the best balance. And the contingent generalizations that figure in the deductive closure of the axiomatic system which strikes the best balance are the laws of nature.[6]

The extent to which we can axiomatize the particular matters of fact depends on how regular our world is. In nice deterministic worlds, we can in principle axiomatize to such an extent that we only need a list of initial conditions under the "facts" heading. At nasty, irregular worlds, only some very small proportion of the particular matters of fact might be axiomatizable, so there won't be very much under the "axioms" heading, and there'll be quite a lot of particular matters of fact left over under the "facts" heading. A regular but indeterministic world will, I suppose, fall somewhere between these two extremes.

Now, why does this get to count as a *prima facie* plausible analysis of lawhood? Answer: because it seems to preserve a good number of our intuitions about what a theory of lawhood ought to do. I'll run through three of the most important.[7]

First up is the thought that any adequate analysis of laws of nature has to distinguish between laws and accidents: between generalizations that are true as a matter of law and those which merely happen to be true. The Ramsey-Lewis view makes this distinction on the basis of whether or not the generalization in question figures as an axiom or theorem in the best system. Consider, for example, the true generalization that everyone currently in Seminar Room E is a philosopher. This is not a law of nature. According to the Ramsey-Lewis view, the reason why it isn't is that it does not figure as an axiom or a theorem in the best deductive system. Adding this generalization to our system—or adding some other axiom that entails it—would yield hardly any gain in strength but would detract significantly from simplicity, so the generalization is not a law.

Next up is the fact that laws support counterfactuals, whereas accidental generalizations do not. This fact is nicely accommodated by Lewis's theory of counterfactuals. According to Lewis, the extent to which the laws are the same is an important feature in determining how similar worlds are to each other. So for example it ought to come out true that if I were to drop my pen now, it would fall with an acceleration equal to the force exerted on it divided by its mass. And on Lewis's analysis this *does* come out as true. We are required to hold the laws fixed as far as possible when looking for the closest world where I drop my pen. We might need a small miracle to get me to drop it in the first place, but we don't need to tinker with the law that $f = ma$. So the closest world where I drop my pen will be one where it is a law that $f = ma$; hence at that world my pen falls with the appropriate acceleration and the counterfactual is true.[8]

Accidental generalizations, on the other hand—like the generalization that everyone currently in Room E is a philosopher—do not in general support counterfactuals on Lewis's account. It's not true, for instance, that if Gavin (who is not a philosopher) were to walk into Room E now he would become a philosopher, since we are not required to hold the truth of the generalization that everyone there is a philosopher fixed when looking for the closest world where Gavin walks in. Thus a world where he enters and retains his status as a non-philosopher is closer than any world where he walks in and his brain bizarrely reconfigures itself in a way which makes him count as a philosopher. (Obviously being a philosopher isn't just to do with what state your brain is in. But whatever it takes to be a philosopher, it won't happen to Gavin in the closest world where he enters the room.)[9]

Third and finally comes the connection between laws of nature and physical necessity: "It is a law that A" is supposed to entail "it is physically necessary that A." Lewis gets this right by means of stipulative definition: A is defined to be physically necessary if and only if it follows from the laws of nature.

Armstrong's View

Those are the three respects in which the Ramsey-Lewis view preserves our intuitions about laws of nature that are going to be most important in what follows. Just for the sake of comparison, I'll give a rough account of how Armstrong's view does the same thing.

For Armstrong, laws of nature are necessary relations between universals. He writes this "$N(F, G)$": the second-order relation N ("N" for "necessitation") holds between first-order universals F and G. According to Armstrong, the holding of N *entails* the generalization that all Fs are Gs: since F-ness necessitates G-ness, all Fs are going to turn out to be Gs.

(In fact, Armstrong thinks that the entailment from $N(F, G)$ to "all Fs are Gs" does not go through for all deterministic laws. It might be that $N(F, G)$ holds but only for those Fs that are not Hs: Fs that *are* Hs are not Gs. Hence

N(*F, G*) does not entail that all *F*s are *G*s. Armstrong calls laws like this "oaken laws," to be contrasted with "iron laws" for which the entailment does hold.[10] Thus when I talk about the entailment between *N*(*F, G*) and "all *F*s are *G*s," the reader should take it that I am talking about iron laws. Note that the iron/oaken distinction does not apply to the Ramsey-Lewis view, according to which *all* deterministic laws are true universal generalizations.)

The distinction between laws and accidents, then, is simply a matter of whether or not the relevant universals are related by *N*. There is a necessary connection between force, mass, and acceleration, but not between being in this room now and being a philosopher. *N* provides the ground for counterfactuals too: if *N*(*F, G*) holds, then, Armstrong says, there is every reason to suppose that it would continue to hold under the counterfactual supposition that some object is *F;* hence that object would be *G* too, as required.[11] Finally the connection between lawhood and necessity is pretty obvious: laws just *are* physically necessary relations.

So those are the two competing views. The Ramsey-Lewis view is (or at least I'm construing it as) a reductionist view, and therefore one according to which the supervenience of laws on non-laws holds at all possible worlds: if two worlds are identical with respect to their particular matters of fact, those facts will have the same best axiomatization and hence the worlds will have the same laws. Armstrong's realist view does not respect supervenience: two worlds can be identical with respect to all their particular matters of fact and yet differ with respect to which universals bear *N* to each other. Indeed it's part of Armstrong's view that there is a world identical to ours with respect to all its non-nomic facts but which has no laws of nature at all—just accidentally true generalizations.[12]

III. Descriptive vs. Governing Conceptions of Laws

I just said that both the Ramsey-Lewis view and the realist view preserve common-sense intuitions about laws and accidents, about the counterfactual-supporting nature of laws, and about physical necessity. What I want to do now is draw attention to the main difference between them. The fundamental difference. I think, is that for the Humean, laws are purely descriptive of the particular matters of fact, whereas for the anti-Humean, laws *govern* what the particular matters of fact are.[13]

One way of bringing this out is to consider the thesis of determinism.

We can characterize determinism in the following rough and ready way: the state of the universe at any given time together with the laws of nature determines what the state of the universe will be at any future time. But what does "determines" mean here? For the Humean, the laws and current facts determine the future facts in a purely logical way: you can *deduce* future facts from current facts plus the laws. And this is just because laws *are,* in part, facts about the future. So for the Humean, the notion of determination is, as it were, a metaphysically thin one. This contrasts sharply, I think, with the notion of determination which the anti-Humean has. For the anti-Humean, the notion of determination is a metaphysically meaty one. It isn't just that the laws plus current facts *entail* future facts; rather the laws "make" the future facts be the way they will be: the laws are the ontological *ground* of the future facts.

Another way of putting this is to say that for the anti-Humean the laws are already present in the current state of the universe. Imagine Armstrong writing down everything that's true of the universe up to this moment. One of the things that will appear in his list will be the obtaining of *N* between various pairs of universals. And it's in the nature of *N* that its obtaining entails that those universals will carry on occurring together. "*N*(*F*, *G*)" expresses a relation that is already with us, so the future really is determined by some current feature of the universe. For the Humean, on the other hand, a complete list of everything that's true of the universe up to now entails nothing whatever about the future, since if future facts by definition are banned from the list, then so are laws of nature. Among the current facts will be true generalizations about what's happened up to now, but none about what's *going* to happen. The current state of the universe in and of itself does not, as it were, contain the seeds of the future.

Still another way of bringing out this fundamental clash of intuitions is to consider one formulation of the problem of free will. One way of putting the main intuition behind incompatibilism is this: If determinism is true, then, given the laws and the current state of the universe, I could not have acted otherwise than the way I did act. Since I could not have acted otherwise, my act was not free.

Now, the premise of the argument is obviously true: if determinism is true, then, given the laws and the current state of the universe, I could not have acted otherwise than the way I did act. But I think a Humean about laws of nature ought to question why this premise is supposed to entail that we are not free. Suppose the act in question is the raising of my arm.

Suppose further (rather implausibly) that the relevant law is that every person in state *P* raises their arm, and that I am in state *P*. It follows that I will raise my arm. It doesn't follow, I think, that I am not free. For on the Humean conception of lawhood, its being a law that everyone in state *P* raises their arm depends upon what happens in the world—and in particular upon whether or not I, who am in state *P*, raise my arm. Given that I really am in state *P*, to say that it is a law that everyone in state *P* raises their arm is already to presuppose that I'll raise my arm; and the sense in which I am thereby constrained to raise it is a purely logical one. And this logical sense surely cannot be an obstacle to free will. Whether or not determinism is true, and however much free will we have, we logically cannot make what will happen fail to happen. If *P* will be true, nobody—not even God—can bring it about that not-*P*. *Qué serà serà* is a logical truth and therefore not something with profound metaphysical implications.

I think those who are moved by this argument for incompatibilism are implicitly adopting an anti-Humean conception of laws of nature—and in particular, a conception of laws according to which laws are not just generalizations about what has happened and will happen, but rather *govern* what will happen. It is this thought which prompts one to think that the laws of nature place a constraint on our actions that is in some way incompatible with freedom: a constraint which forces us in some metaphysical, not-purely-logical sense to act in the way we do.[14]

This detour into determinism and free will is a way of making vivid the profound difference between seeing laws as descriptive, as the Humean does, and seeing them as governing, as the anti-Humean does. For the Humean, since the laws are descriptive, what the laws are depends on what the facts are—including future facts. For the anti-Humean, what the facts are depends upon what the laws are. Humeans and anti-Humeans therefore have completely opposite conceptions of what provides the metaphysical basis for what.

The Humean and anti-Humean positions, then, differ radically with respect to the roles that they expect laws of nature to perform; and it is these different expectations that give rise to the differences in ontology. For the anti-Humean, laws (unlike accidentally true generalizations) *do* something—they *govern* what goes on in the universe—and they therefore require some sort of ontological basis (*N*, for instance) that gives them this ability. Humeans, on the other hand, do not require laws to "do" anything: like accidentally true generalizations, laws are at bottom merely true descriptions of what goes on. Thus for the Humean there is no need for any ontological distinction between laws and accidents.

IV. It Is Not a Conceptual Truth That Laws Govern

I would guess that unreflective common sense favors the governing conception of laws over the descriptive conception: it appears to be natural and tempting to think that the laws of nature determine how things behave in the metaphysically meaty, anti-Humean sense. It would also be natural and tempting for the anti-Humean to try to use this fact about the common-sense conception of lawhood as the basis of a quick and devastating argument against Humeanism: if it is part of the concept of lawhood that laws of nature play a governing role, and Humeanism accords them no such role, then Humeanism about laws must be false. According to this line of argument, Humeanism is based on a conceptual error: that of thinking that it is conceptually possible for something which does not govern to be a law of nature.

As far as I know, nobody has ever used such a blunt and direct argument against Humeanism. But it will become evident in sections V and VI below that some purported counter-examples to Humeanism are essentially rather heavily disguised versions of the argument. So if it can be shown that the argument does not succeed in its blunt form, then the counter-examples do not succeed either.

In what, then, does the alleged conceptual connection between lawhood and government have its roots? The only answer that suggests itself lies in the conceptual connection which undoubtedly does exist, at least in some cases, between the notion of law and the notion of a law*giver:* some being, institution, or other authority (Her Majesty's Government, the MCC, or God, for instance) whose decrees constitute the laws of the land, the rules of cricket, or whatever. Given the naturalness of the connection between laws and a lawgiver, it is natural to think of laws of nature in the same light: as decrees, rather like the Ten Commandments, laid down at the beginning of the universe, to be obeyed without exception by everything from the smallest particle to the largest galaxy. The idea that the laws of nature represent something like God's cosmic plan for the universe is an old one; and it is a view which has undoubtedly survived, at least in some quarters, to the present day—a fact to which a glance at the popular science titles in any bookshop will testify.[15] Still, it is one thing to claim that the laws of nature represent a piece of divine legislation; it is quite another to think that we are conceptually *required* to think of laws of nature in this way. I take it to be just plainly true that belief in laws of nature does not conceptually presuppose belief in a divine lawgiver; hence if it is supposed to be a conceptual truth that laws govern, then we must look elsewhere for a vindication of the alleged conceptual connection between the notion of lawhood and the notion of government.

Perhaps a pertinent analogy is to be found between laws of nature and moral laws. We take it to be conceptually necessary that moral laws have some sort of jurisdiction over us; but we do not take it that the idea of a moral law conceptually presupposes that there is an author of those laws. So—given that we ought not to think that the idea of a law of nature presupposes that there is an author either—it might still be regarded as conceptually necessary that laws of nature have some sort of jurisdiction too.

Is there any sound conceptual basis for taking this analogy with moral laws seriously enough to merit the claim that it is a conceptual truth that laws of nature govern? I think not; for there is an important difference between the ways in which moral (and other) laws on the one hand, and laws of nature on the other, are supposed to have jurisdiction over the entities which fall within their scope; and this difference undermines the analogy. The difference is that moral (and other) laws are *prescriptive,* whereas laws of nature are supposed to "govern"; and whatever governing amounts to in this context, it is a very different kind of jurisdiction than that of prescriptive laws.

For one thing, prescriptive laws only establish how agents *ought* to behave—they do not logically constrain agents to behave in accordance with them—whereas laws of nature *do* logically constrain objects to behave in accordance with them. "Thou shall not steal" is not at all like "thou shalt not transmit signals faster than the speed of light." People do violate moral laws; in fact it seems to me to be essential to the idea of a moral law that it is breakable. It's hard to see how something could count as a moral law—or a rule of cricket, or whatever—if nobody was capable of breaking it. Consider, for example, the absurdity of the idea of having a rule of cricket which states that nobody may bowl faster than 500 miles per hour. Since human beings are physiologically incapable of bowling that fast, it's not clear how there could be such a rule, since it is a rule that is incapable of influencing anyone's behavior. Laws of nature, by contrast with other kinds of law or rule, cannot be broken, since they *do* entail true generalizations. If it is essential to the prescriptive force of other kinds of law that they can be broken, and laws of nature cannot be broken, then laws of nature are unsuited to being cast in a way that is supposed to be analogous to the prescriptive nature of other laws.

For another thing, laws and rules in general, with the exception of laws of nature, govern behavior by means of imposing sanctions on those who do not obey them: you go to Hell, suffer from guilt, get slung in jail, or sent off the pitch. Even if nobody ever violates a law, part of what gives the law its prescriptive force is the fact that if you were to break it something bad would, or might, happen to you as a result. Laws without the threat of sanctions again seem not to make much sense; for instance I can't imagine a world where there is a rule banning handling the ball in soccer but where there is no threat of punishment for offenders. For in such a world the rational soccer player will handle the ball whenever it's in his interests to do so, and the game will end up indistinguishable from the way it's played in a world where there is no such rule. So the alleged "rule" doesn't seem to be doing anything and hence seems not to deserve to be called a genuine rule. Again, the notion of sanction makes no sense when applied to laws of nature: it's hardly as if potentially recalcitrant objects are kept in line by the threat of punishment. Hence, again, no analogy seems possible between laws of nature and other kinds of law.

What does all this show? Well, it is very plausible to think that if common sense *does* take it to be part of the concept of a law of nature that those laws govern, then it does so only because of a tacit assumption that laws of nature operate in a way that is analogous to the way that other laws—laws which really do govern—operate. But, as I have argued, that assumption cannot be maintained, since the alleged governing nature of natural laws would have to be entirely unlike the prescriptive nature of moral and other laws. In other words, it would be highly *im*plausible to maintain that a conceptual connection can be traced between the concept of a law of nature and the concept of government that is independent of that assumption. *A priori* reflection on the nature of natural laws by themselves does not yield any requirement to think of them as playing a governing role.

This is not, of course, to say that it is merely a linguistic accident that laws of nature came to be so called. On the contrary: I dare say that the term was introduced in order to capture the belief that what happens in the universe happens as a matter of divine decree—in which case the expression "law of nature" really did start out with an explicitly legislative connotation.[16] But the cultural context in which the expression was coined is not ours. If we are to use the expression, as we do, in a way that is consistent with the laws of nature being nobody's decree, then we thereby lose any presumption in favor of regarding them as having any kind of jurisdiction over what happens in the universe.

Of course, none of this establishes that there is anything incoherent, or even wrong, in taking laws of nature to govern. For all I have said, Armstrong may yet be right that laws are necessary relations between universals, and hence right to think that laws determine, in a metaphysically substantive sense, what happens. All I hope to have established is that we are not conceptually *required* to think that just because laws of nature are called "laws," they must play a governing role. Hence it is no objection to Humeanism to complain that it accords laws of nature no such role: the intuition that laws of nature govern is an optional one. And indeed part of the motivation for Humeanism about laws is the desire to steer well clear of the sorts of ontological commitments that are needed to shore up the intuition that laws govern.

So far, then, I hope to have shown that the Ramsey-Lewis view provides a conception of laws that is a well-motivated and at least *prima facie* plausible alternative to the anti-Humean conception, since there are no persuasive reasons to think that there is something conceptually awry in the claim that laws do not govern. In sections V and VI, I shall discuss two thought experiments that are designed to show that Humeanism about laws—and *a fortiori* the Ramsey-Lewis view—is wrong: wrong because we can imagine possible situations where the verdict of Humeanism is not the verdict delivered by pre-theoretical, common-sense intuition.

The reason why the thought experiments fail to establish this, I think, is that they do not address the Humean about laws on her own terms, or even on neutral terms, but rather presuppose that laws have certain features which the Humean unashamedly rejects—or at least, ought to reject. In particular the common-sense intuitions which the thought experiments appeal to are ones that involve an implicit commitment to the view that laws govern. If what I've said so far is right, then the intuition that laws govern is an optional one: we need not think of laws that way. Hence the intuitions which the thought experiments try to bring to bear against Humeanism are optional too.

V. *X*-Particles and *Y*-Fields

The first counter-example has been defended by Tooley, by Carroll, and by Peter Menzies; I'm using Carroll's formulation.[17] It involves a possible world, w_1, consisting entirely of *X*-particles and *Y*-fields. No *X*-particle ever gets to enter a *Y*-field. But according to Carroll it might nonetheless be a law that all *X*-particles in *Y*-fields have spin up—and indeed that this might be the *only* law at w_1. (Call this law L_1.) Now consider w_2, which is identical to w_1 with respect to particular matters of fact but which allegedly differs nomically: at w_2 it is a law (call it L_2) that all *X*-particles in *Y*-fields have spin *down*. Since w_1 and w_2 differ with respect to their laws of nature but not with respect to their particular matters of fact, laws cannot supervene on non-laws: the Humean conception of laws, and *a fortiori* the Ramsey-Lewis view, is false.

The only real option for the upholder of the Ramsey-Lewis view (or at least the strong version of it, according to which nomic facts *logically* supervene on particular matters of fact) is, as Carroll points out, to deny that w_1 and w_2 are possible worlds at all. And he claims that the only grounds on which the Humean could base this claim would be to deny that there could be a world with only one basic law, or to deny that there can be vacuous laws—or both.[18]

Well, I'm not sure if there could be a world with only one basic law, but I'm happy to concede that there could be. And the Ramsey-Lewis view certainly allows that there can be vacuous laws. So if these are the only legitimate grounds for denying that w_1 and w_2 are possible, as Carroll claims, then Humeans are in trouble. Luckily they aren't the only legitimate grounds. For note that Carroll's counter-example does *not* show that the Ramsey-Lewis view fails by *its own lights:* it isn't as if the Ramsey-Lewis view *entails* that w_1 and w_2 are possible, thus undermining its own claim to respect the supervenience of laws of nature on particular matters of fact. The Ramsey-Lewis view *itself* judges w_1 and w_2 to be impossible. According to the Ramsey-Lewis view, if L_1 is to count as a genuine law in w_1 then it must appear in the axiomatization of the best true theory of w_1. But it doesn't. Without L_1 we have—by Carroll's stipulation—no laws, and hence no axioms, at all. *With* L_1 we have a single law which contributes no positive virtue whatever, since—being vacuous—it does not systematize a single fact. So the Ramsey-Lewis view dictates that L_1 is not a law in w_1, and similarly that L_2 is not a law in w_2. In other words, according to the Ramsey-Lewis view w_1 and w_2 are not possible worlds at all.

Now, all I've really done here is relocate the problem. Carroll will say, I suppose, that the Ramsey-Lewis view renders w_1 and w_2 impossible whereas common-sense intuition says they are possible. So the Ramsey-Lewis view violates our intuitions about laws. Well, the Ramsey-Lewis view clearly violates Carroll's intuitions—and indeed the intuitions of anyone who already thinks that laws are somehow "out there," prior to and watching over the particular matters of fact to make sure they don't step out of line. If you think that way, then of course you'll be able to make sense of the idea that w_1 might be "governed" by L_1 even though L_1 never gets instantiated.

But of course the fact that the Ramsey-Lewis view violates *these* sorts of intuition is no cause for alarm, since the whole starting point of the Ramsey-Lewis view in the first place, as I said earlier, was to deny this mysterious supposed "governing" feature of laws. To assert that w_1 and w_2 are possible is, at bottom, merely to assert that laws govern rather than describe; and to deny their possibility is merely to assert the opposite. Carroll's alleged counter-example, then, is really just a restatement of this basic anti-Humean intuition, and as such poses no threat to the Humean.

There are two further points that need to be made about this counter-example. The first is that I'm in no way denying the possibility of vacuous laws *in general*. For instance we *can* imagine two worlds, w_3 and w_4, each of which *contains* the situation described in the counter-example and which have L_1 and L_2 respectively as vacuous laws. Suppose, for example, that in w_3 there are lots of *other* kinds of particle which *do* get to enter Y-fields, and acquire spin up when they do so. And suppose that X-particles are sufficiently similar to those other kinds of particle to make the best system of generalizations one which does not regard X-particles as a special case. In other words, the best system of axioms might have it that *all* particles in Y-fields have spin up, and thus have it as a theorem that all X-particles in Y-fields have spin up too. Similarly for w_4 with respect to L_2. So I'm not ruling out vacuous laws *tout court;* only ones whose inclusion in the best system is not warranted by some improvement in that system.

The second point is that there is an important lesson to be learned from the counter-example's failure—namely that we ought to be very cautious about the extent to which we trust our pre-theoretical intuitions about laws, since to take Carroll's counter-example seriously commits one to a *very* strong realism about laws. The counter-example is incompatible not just with the Ramsey-Lewis view, but also with Armstrong's realism—since by hypothesis L_1 and L_2 are laws which "govern" properties (namely the property of being an *X*-particle in a *Y*-field and the property of having spin up) which are never instantiated. And according to Armstrong's theory of universals there are no uninstantiated properties. For Armstrong, then, w_1 and w_2 are impossible because *N*—the law relation which holds between universals—is not instantiated at either world. So commitment to the possibility of w_1 and w_2 commits one either to the view that w_1 and w_2 contain universals that are never instantiated, or to some other realist analysis of lawhood according to which laws are not relations between universals but are rather some other kind of entity capable of governing. Each option requires ontological commitments which even an anti-Humean ought to regard with a good deal of suspicion, not least because it is very hard to imagine how the postulated entities might get to perform the governing job required of them.

VI. The Mirror Argument

The "Mirror Argument" has a fairly central place in Carroll's book on laws of nature.[19] In common with most thought experiments that are designed to scupper Humeanism about laws, it involves a dull, barren, and very distant possible world, consisting only of our old friends the *X*-particles and *Y*-fields, together with a mirror on a swivel. It goes like this:

Consider a possible world, U_1, consisting of exactly five *X*-particles, five *Y*-fields, and not much else. Each of the five particles enters a *Y*-field at some point, and each particle acquires spin up when it does so. The particles all move in a straight line throughout all eternity. But close to the route of one of the particles—call it particle *b*—there is a mirror on a swivel. In fact, the mirror is in a position (position *c*) such that it does not interfere with the trajectory of particle *b:* the particle just passes right on by. But had the mirror been swiveled round to position *d* (or if it had just always been in position *d*), it would have been right in the path of particle *b,* and the particle would have been deflected away from its *Y*-field. Finally, let us suppose that it is a law in U_1 that all *X*-particles subject to a *Y*-field have spin up. Call the statement that all *X*-particles subject to a *Y*-field have spin up *L;* so *L* is a law in U_1.

Now consider possible world U_2. U_2 is just the same as U_1 except that in U_2 particle *b* does not acquire spin up when it enters the *Y*-field. So *L* is not a law at U_2 because it isn't even true: not all X-particles in Y-fields in U_2 have spin up. It's important to note, however, that the recalcitrant particle b does not differ in respect of its natural, intrinsic properties from all the other X-particles; it's no part of the story that there's some sort of *explanation* of why b behaves bizarrely at U_2 but not at U_1.

Now, how does this set-up tell against Humeanism about laws? Well, the alleged problem comes not from U_1 and U_2 themselves, but from considering what *would* have happened in each of U_1 and U_2 had the mirror been in position d rather than position c—that is to say, if the position of the mirror had stopped particle b from entering a Y-field. In particular, we're interested in what the *laws* would have been had the mirror at U_1 and U_2 respectively been in position d.

To find this out, we have to look at the closest possible world to U_1 where the mirror is in position d—call this world U_1*—and the closest possible world to U_2 where the mirror is in position d; call this world U_2*. And we have to see what the laws are in each of U_1* and U_2*.

First, U_1*. Remember that L is a law at U_1; particle b acquires spin up when it enters a Y-field. So it seems eminently reasonable to suppose that L is a law at U_1* too—since U_1* only differs from U_1 in that particle b never finds its way into a Y-field. Now U_2*. Remember that L isn't true—and hence not a law—at U_2, since at U_2 particle b fails to acquire spin up when it enters a Y-field. But at U_2*, L *is* true, since our troublesome particle b never gets to exhibit any troublesome behavior. Intuitively, though, according to Carroll, although L is *true* at U_2* it is only accidentally true: it is not a *law* at U_2* that all X-particles in Y-fields have spin up. L only gets to be true at U_2* in virtue of the purely contingent, fortuitous positioning of the mirror. But U_1* and U_2* are the same with respect to their particular matters of fact, yet they differ with respect to whether or not L is a law. So laws do not logically supervene on particular matters of fact.

Now, Carroll in fact offers a formal and, in his view, more persuasive way of getting to the same conclusion, by appealing to a couple of modal principles. I'll deal with the more formal argument first, and then come back to the rather less formal, more intuitive characterization of the counter-example which I just outlined.

So, the formal version first. Carroll offers the following principle, (SC), as an intuitively compelling modal principle (where $\Diamond_P$ and $\Box_P$ mean "physically possible" and "physically necessary" respectively):

(SC) If $\Diamond_p P \& \Box_p (P \supset Q)$, then if *P* were the case, *Q* would be the case.

From this he derives two other principles:

(SC*) If $\Diamond_p P$ and *Q* is a law, then if *P* were the case, *Q* would still be a law.

(SC') If $\Diamond_p P$ and *Q* is not a law, then if *P* were the case, *Q* would still not be a law.

The argument against Humeanism about laws then proceeds like this: Let *P* be the sentence "the mirror is in position *d*." We can suppose that *P* is physically possible at both U_1 and U_2. Hence by (SC*), since *L* is a law at U_1, it's true at U_1 that if *P* were the case, *L* would still be a law. Call the world which makes this counterfactual true—that is, the world closest to U_1 where *P* is true—U_1*. (SC*) entails that *L* is a law at U_1*. Furthermore, by (SC′), it's true at U_2 that if *P* were the case *L* would still *fail* to be a law, since *L* isn't a law at U_2. Call the world which makes *this* counterfactual true—that is, the world closest to U_2 where *P* is true—U_2*. (SC′) entails that *L* is not a law at U_2*. But since U_1* and U_2* are identical with respect to particular matters of fact and differ with respect to whether *L* is a law, laws do not supervene on particular matters of fact.

Doubtless the argument is valid; but in order for it to be persuasive we need to have good reason to believe the premises, (SC*) and (SC′). And one good reason to *doubt* the premises is that according to one popular analysis of counterfactuals, namely Lewis's, they are false.

(SC*) and (SC′) are both false on Lewis's analysis because they both assume that the closest possible world to a given world *w* where *P* is true is a world that is physically possible with respect to *w;* and on Lewis's analysis this does not hold in general. I'll just give an example to show that (SC*) is false; parallel reasoning can be applied to (SC′).

Suppose the actual world is deterministic. Then worlds that are physically possible relative to our world will in general be radically different to ours with respect to their particular matters of fact. For example let

P = Natalie drops her pen at *t*

and suppose that in fact Natalie hangs on to her pen at *t,* so at the actual world, *P* is false.

Now, doubtless there are physically possible worlds where *P* is true. Call the closest such physically possible world *w**. *w** must differ from ours in at least some respect for all times before *t*, since if it were identical to ours at some time before *t* it would also be identical to ours for all future times, including *t* itself; in which case *P* would be false at *w** too, which, by assumption, it isn't. So despite being the same as our world as far as its laws are concerned, *w** is a pretty distant world, since it doesn't ever have perfect match of particular matters of fact with our world.

Now consider not the closest *physically* possible world where *P* is true, but just the closest possible world where *P* is true. Call this world w_p. On Lewis's analysis, this will be a world which is identical to ours with respect to particular matters of fact until just before *t;* but in w_p there is a small "miracle" just before *t* so that *its* laws (unlike ours) allow Natalie to drop her pen. Now let *Q* be whichever law of ours it is that's broken at w_p.

Relative to the actual world, *P* is physically possible and *Q* is a law. So the antecedent of (SC*) is true. But the consequent is false: it's false that if *P* had been the case, *Q* would still have been a law. So (SC*) is false, because the closest possible world where *P* is true need not be one that is physically possible: *w** and w_p are different possible worlds.

Thus the modal principles which form the premises of the Mirror Argument are false according to Lewis's analysis of counterfactuals. Still, one might want to claim that the modal principle (SC), from which (SC*) and (SC′) are derived, is intuitively compelling; hence we should take its verdicts, rather than those given by Lewis's analysis, to be true. This is certainly Carroll's view. In defense of (SC) he asks us to consider the counterfactual "if the match were struck, then the laws would be different." After noting that this counterfactual is true according to Lewis's analysis, he says, "there are plenty of reasons not to abandon (SC) or (SC*) . . . most importantly, it is obvious that the laws do not counterfactually depend on the striking of a match. If the match were struck, the laws would be no different. That is *so* obvious that I have trouble believing that anyone, especially Lewis, would hold that some of our laws would not still be laws if that match were struck."[20]

In the light of the discussion in section III, this passage should strike the reader as one with a distinctly anti-Humean flavor: the view that "it is obvious that the laws do not counterfactually depend on the striking of a match" is clearly born of the view that particular matters of fact depend upon the laws and not vice versa. If the only support to be had for (SC) is such an explicit avowal of the anti-Humean conception of laws, then the Mirror Argument is clearly question-begging.[21]

However, Carroll does offer a different defense of (SC) near the beginning of *Laws of Nature,* where he argues that (SC) is a plausible consequence of his "picture" of laws of nature. "Support for this principle [(SC)] comes," he says, "from a familiar picture of reality which embodies especially vividly the concept of lawhood employed in common sense."[22]

And the picture looks like this: "the view of laws as the edicts of a lawgiver does provide a useful metaphor. I rely on this metaphor insofar as it underlies a more secular and more detailed picture: *the Laplacean picture.* This worldview includes a portrayal of our universe as completely determined by its temporally local history at any one time together with a statement of what propositions are laws. According to the Laplacean picture, it is *as if* God created the world by designating the initial conditions and the laws."[23]

Well, this is indeed a familiar picture; but of course it's just this picture—the "Laplacean creation myth," as Loewer[24] calls it, which lies behind the anti-Humean conception of laws. As I argued in section III, and contrary to what Carroll supposes, there is an enormous difference between thinking merely that the universe is "completely determined by its temporally local history at any one time together with a statement of what propositions are laws" and thinking that "it is *as if* God created the world by designating the initial conditions and the laws." By buying into the Laplacean creation myth at the outset, Carroll is simply justifying (SC) by appealing to a conception of laws which the Humean explicitly repudiates. Thus no Humean ought to be persuaded by Carroll's justification of (SC); hence the formal version of the Mirror Argument begs the question against the Humean.

This leaves the rather more intuitive informal version of the argument. This differs from the formal argument in that rather than deriving *L*'s lawful status at U_1* but not U_2* from precise modal principles, it invites us to consult our *intuitions* about its lawful status at each world. And intuition, Carroll thinks, delivers the same result: *L* is a law at U_1* but merely an accident at U_2*; hence Humeanism about laws is false.

Recall that U_2 is the world just like U_1 except for its rogue particle which fails to acquire spin up as it goes through the *Y*-field. U_2* is the closest world to U_2 where the mirror is in position *d,* so that the rogue particle doesn't get to pass through the *Y*-field at all. The intuition, then, is that it's just an accidental feature of U_2* that *L* is true; hence *L* is not a law there.

Carroll locates the source of this intuition in the following thought: "it is natural," he says, "to think that *L*'s status as a law in U_1 does not depend on the fact that the mirror is in position *c* rather than position *d* . . . it is just as natural to think that *L*'s status as a non-law in U_2 also does not depend on the position of the mirror. *L* would not be a law in U_2 even if the mirror had been in position *d* . . . The question the friends of supervenience must face is how they are going to ground the fact that *L* is a law in [U_1*] but not [in U_2*]" (1994, p. 62).

Well, as a friend of supervenience, I have no desire to find a way of grounding the "fact" that *L* is a law in U_1* but not in U_2*, since I think *L* is a law in U_2* and not an accident. This commits me to the apparently unacceptable claim that the position of the mirror in U_2 affects what the laws of nature are, since I am committed to the truth of the counterfactual "if the mirror had been in position *d, L* would have been a law." But I truly see no harm in that. Recall that in our world, what the laws of nature are depends upon whether or not Natalie drops her pen in just this counterfactual sense (since, as we've seen, if Natalie *had* dropped her pen, the laws of nature would have been different). As I said earlier, part of the Humean creed is that laws of nature depend on particular matters of fact and not the other way around; it is no surprise to the Humean, then, that by counterfactually supposing the particular matters of fact to be different one might easily change what the laws of nature are too.

The intuition that's really doing the work in this counter-example, then, is the intuition that laws are not purely descriptive: the thought being that since particle *b*'s behavior is not *governed* by any law at U_2, it can't be governed by a law at U_2* either. Hence *L* can't be a law at U_2*, and hence Humeanism is false. But to describe the example in those terms is not to describe it in neutral terms but to describe it in terms which explicitly presuppose an anti-Humean starting point. Indeed I don't think there's any way of trying to motivate the intuition that *L* is not a law in U_2* that *doesn't* presuppose an anti-Humean starting point, either by appealing to the notion of government, or to a modal principle like (SC), or by making the claim that the laws do not depend on the facts. On the other hand, if you start off by buying into the story I told earlier about the descriptive conception of laws, I can't see how Carroll's thought experiment could possibly move you to say that *L* isn't a law at U_2.

One reason I may not have convinced you yet is that you may be tempted by the following thought: there is an obvious *counterfactual* difference between U_1* and U_2*; hence they cannot have the same laws, and therefore cannot be the same world—contrary to what I am insisting. And the counterfactual difference you might point to is this: at U_1*, if the mirror had been in position *c,* particle *b* would have had spin up. But at U_2*, if the mirror had been in position *c,* particle *b* would *not* have had spin up. This seems to be a way of explaining the intuition that *L* is not a law at U_2 which does not explicitly make reference to any off-limits anti-Humean views: it just appeals to nice, neutral counterfactual intuitions.

Well, it does on the surface. But I deny that there is such a counterfactual difference between U_1* and U_2*. On what basis, after all, might one claim that the counterfactual difference exists? I can think of three options, and none of them succeeds.

First, you might have a worry which goes something like this: Look, particle *b* at U_2 is a really weird particle, since it doesn't do what all the other *X*-particles do. When you get to U_2* by flipping the mirror, you stop particle *b* from displaying any aberrant behavior. But it's still the same particle, so it must somehow be in *b*'s nature not to be *disposed* to acquire spin up in a *Y*-field. So it's still true at U_2* that if particle *b were* to go through a *Y*-field, it would fail to get spin up; hence the counterfactual difference between U_1* and U_2* really does obtain, since particle *b* at U_1* *is* disposed to acquire spin up in the presence of a *Y*-field—although of course this is not a disposition which it ever gets to manifest.

But this appeal to the difference between *b*'s alleged dispositional properties in different worlds doesn't work. By stipulation, U_1* and U_2* are identical with respect to their particular matters of fact. *A fortiori* there is no difference in intrinsic non-nomic properties between U_1*, where *b* is disposed to acquire spin up, and U_2*, where it isn't. What, then, is the difference in dispositional properties supposed to amount to? You can't say that the U_1* version of *b* has a different disposition to the U_2* version of *b* because of any difference in their non-nomic properties, since by hypothesis there are no such differences. Nor can you say that the U_1* version of *b* has a different disposition to the U_2* version of *b* because the laws of nature are different at U_2*, since the claim that they have different laws is precisely the conclusion you're trying to motivate. So the appeal to dispositions fails.

Second, you might be convinced by the following argument. The mirror is in position *c* at U_2. U_2* is the closest possible world to U_2 where the mirror is in position *d.* So U_2 must be the closest possible world to U_2* where the mirror is in position *c*. Therefore at U_2* it must be true that if the mirror had been in position *c,* particle *b* would have had spin up. Luckily this is a fallacious piece of reasoning. Just because it's true at U_2 that U_2* is the closest possible world where the mirror is in position *d*, it doesn't follow that it's true at U_2* that U_2 is the closest possible world where the mirror is in position *c*. There might easily be other worlds where the mirror is in position *c* that are closer to U_2* than U_2 is; U_1 for instance.

You might still want to say that the alleged counterfactual difference between U_1* and U_2* exists. But—and this is the third attempt—in that case you need to find an analysis of counterfactuals which backs you up. However, since analyses of counterfactuals typically require that the laws of nature be held fixed as far as possible when looking for the closest world where the counterfactual's antecedent is true, the only way you're going to be able to make it true at U_2* that if the mirror had been in position *c*, particle *b* would not have had spin up is by presupposing that *L* is not a law at U_2*. And since *L*'s status at U_2* is precisely what's under dispute, you can't presuppose that it's not a law without begging the question.

So I think there's absolutely nothing wrong with saying that *L* is a law at U_2*; hence the counter-example fails. I should add that I don't at all mean to suggest that there's anything *incoherent* about thinking that U_1* and U_2* have different laws. I just think that if you *have* that intuition, you aren't going to be able to spell out *why* it's a plausible intuition without appealing to the sorts of anti-Humean assumptions which by definition the Humean isn't the least bit interested in accommodating.

VII. Conclusion

Humeanism about laws—and the Ramsey-Lewis view in particular—is a well motivated view, and an attractive one for those driven by either a desire for ontological economy or the Humean suspicion that necessary connections are unintelligible. But it's not a position to be taken on board lightly. If you want to be a Real Humean—one who is genuinely unmoved by counter-examples like Carroll's Mirror Argument—you have to purge a lot of intuitions about laws that are quite widely accepted both in the philosophical literature and among the rather less philosophically reflective.

But I still maintain that the Ramsey-Lewis view does justice to enough of our intuitions about the role of laws of nature to be a viable alternative to Armstrong's realism. I suspect this is a claim which some anti-Humeans will deny, on the grounds that while the Ramsey-Lewis view does justice on a superficial level to the "gross" intuitions about laws mentioned in section II, it does not do justice to the deeper intuition that there must be some sort of fundamental ontological grounding for the gross intuitions. Thus, with respect to the law/accident distinction, Armstrong says "My sense of the matter is that there is a huge ontological gulf here [between laws and accidents], where a regularity theory can find only a relatively small distinction.[25]

The anti-Humean might try (though Armstrong does not explicitly do so) to claim that it is *these* intuitions about the nature of laws—that is, intuitions which demand some sort of ontological *grounding* for the law/accident distinction and so on—which need to be satisfied by any adequate theory of laws. But of course the Humean response is to say that there *is* no feature of the world which can satisfy *those* intuitions; if that's what something has to do in order to count as a law of nature, then there are no laws of nature.

We can either conclude that there are no laws, or abandon some of our rather more controversial intuitions about what laws are like, and proceed with a rather less metaphysically rich account of lawhood. Part of the Humean claim is that even if we eschew the notion that laws govern, we are still left with things which deserve the title "law of nature"—although perhaps, given the connotations of "law" discussed in section IV, it would have been better if laws of nature had been given a rather less suggestive name.

Notes

This paper was written while I was a Postdoctoral Fellow at the Philosophy Program in the Research School of Social Sciences, Australian National University. I thank the staff and students there for helpful discussion. Thanks are also due, for criticism of earlier drafts, to David Armstrong, Henry Fitzgerald, Michael Smith, Nick Zangwill, and two anonymous referees for *Philosophy and Phenomenological Research*.

1. See Armstrong (1983). The view is sometimes called the "Dretske-Tooley-Armstrong view," since Dretske (1977) and Tooley (1977) came up with similar views at the same time as, but independently of, Armstrong. I shall confine myself to Armstrong's formulation of the view since it is the most well-known and detailed of the three.

2. See Ramsey (1978); Lewis (1973), pp. 73–75; and Lewis (1986), Postscript C. The Ramsey-Lewis view is sometimes known as the Mill-Ramsey-Lewis view, in recognition of the fact that it originated with Mill; see Mill (1875), Book III, Chapter IV.

3. Throughout the paper, the expression "particular matters of fact" is taken to cover facts about objects and their natural, *non-nomic* properties and relations. I leave it as an open question whether laws of nature obey the stronger condition which Lewis calls "Humean supervenience," according to which laws of nature (*inter alia*) supervene on the arrangement of qualities of point-sized entities (see the Introduction to Lewis [1986a]).

4. See van Fraassen (1989), Chapter 3; Carroll (1990); Carroll (1994), Chapter 3; and Tooley (1977).

5. Of course, simplicity isn't really that simple, for there is also the question of how simple the axioms individually are, which I take to be a matter of their logical form and the sorts of predicates which they employ.

6. If there is a tie for the best system, then the laws of nature are whichever axioms and theorems appear in all the tied best systems (see Lewis 1986, 124).

7. A more comprehensive list of intuitions about lawhood, together with a discussion of which of these beliefs are accommodated by the Ramsey-Lewis view, is to be found in Loewer (1996), section IV. Loewer points out, as I do in section III below, that the Ramsey-Lewis view fails to do justice to the intuition that laws govern events.

8. What if we *do* need to tinker with the law that $f = ma$ in order to get me to drop my pen? Then $f = ma$ will not be a law in the closest world where I drop it. Even so, since $f = ma$ is a law at *our* world, we hold it fixed *as far as possible* when determining similarity. We only need the one violation of $f = ma$ in order for me to drop my pen; hence at the closest world where I do so there will only be that one violation. Hence in that world the pen, once dropped, will fall according to $f = ma$.

9. I have appealed to Lewis's theory of counterfactuals here. Of course, in principle one can hold the Ramsey-Lewis view without endorsing Lewis's theory of counterfactuals. My point here is merely to show that *there is* a theory of counterfactuals to which a defender of the Ramsey-Lewis view can appeal in order to do justice to the intuition that laws support counterfactuals.

10. See Armstrong (1983), pp. 147–49. The distinction springs from Armstrong's thesis that there are no negative universals; hence *N* cannot relate, say, *F* and not-*H* with *G*, since one of the alleged relata does not exist.

11. See Armstrong (1983), p. 103.

12. See Armstrong (1983), p. 71.

13. The difference has of course been noticed before; see for instance Swartz (1995) and Loewer (1996). However there has been a tendency among anti-Humeans to ignore the difference and to simply assume that laws' ability to govern is obvious and uncontroversial.

14. The argument that a Humean conception of laws undermines incompatibilism is presented in greater depth in Swartz (1985), chapters 10 and 11, and Berofsky (1987).

15. See Wertheim (1997) for an illuminating popular account of the relationship between science—particularly physics—and religion. Wertheim argues that physicists have always had, and continue to have, a tendency to regard their discipline as a quest for "God's cosmic plan"—the most obvious recent example being Hawking (1988).

16. Indeed, as Ruby (1986, p. 341) notes, thinkers as diverse as Aquinas and Robert Boyle have voiced objections to calling laws of nature "laws" on the grounds that the prescriptive connotations of "law" cannot be applied to the behavior of non-conscious objects. Still, given that their objections went unheeded it seems more sensible to dispute the connotations than to complain that what we call "laws of nature" are not really laws at all.

17. See Tooley (1977), pp. 669–72; Carroll (1990), pp. 202–4; and Menzies (1993), pp. 199–200.

18. See Carroll (1990), p. 203.

19. See Carroll (1994), pp. 57–68.

20. Carroll (1994), pp. 186–87.

21. The issue of what question-begging amounts to is a thorny one. My use of the term here relies on Frank Jackson's analysis, according to which (roughly) an argument is question-begging if the evidence which is adduced in support of the premises of the argument is such that it would not *count* as evidence for a sane person who already doubted the truth of the conclusion. In the present case, the "evidence" adduced by Carroll in support of (SC) would hardly be accepted as evidence by a (sane) Humean who doubted the conclusion of the Mirror Argument. See Jackson (1987), chapter 6.

22. Carroll (1994), p. 17.

23. Ibid.

24. Loewer (1996), p. 115.

25. Armstrong (1993), p. 174.

References

Armstrong, D.M. (1983): *What Is a Law of Nature?* Cambridge: Cambridge University Press.

Armstrong, D. M. (1993): "Reply to Smart," in Bacon, Campbell & Reinhardt (1993), 169–74.

Bacon, J., K. Campbell & L. Reinhardt (eds.) (1993): *Ontology, Causality and Mind.* Cambridge: Cambridge University Press.

Berofsky, B. (1987): *Freedom from Necessity.* London: Routledge & Kegan Paul.

Carroll, J. W. (1990): "The Humean Tradition," *Philosophical Review* 99, 185–219.

Carroll, J. W. (1994): *Laws of Nature.* Cambridge: Cambridge University Press.

Dretske, F. (1977): "Laws of Nature," *Philosophy of Science* 44, 248–68.

Hawking, S. (1988): *A Brief History of Time.* London: Bantam Press.

Jackson, F. (1987): *Conditionals.* Oxford: Basil Blackwell.

Lewis, D. K. (1973): *Counterfactuals.* Cambridge, Mass.: Harvard University Press.

Lewis, D. K. (1986): "Postscripts to 'A Subjectivist's Guide to Objective Chance'" in his (1986a), 114–32.

Lewis, D. K. (1986a): *Philosophical Papers,* vol. II. New York: Oxford University Press.

Loewer, B. (1996): "Humean Supervenience," *Philosophical Topics* 24, 101.

Menzies, P. (1993): "Laws of Nature, Modality and Humean Supervenience," in Bacon, Campbell & Reinhardt (1993).

Mill, J. S. (1875): *A System of Logic.* London: Longmans.

Ramsey, F. P. (1978): "Universals of Law and Fact" in his *Foundations,* ed. D. H. Mellor. London: Routledge & Kegan Paul, 128–32.

Ruby, J. E. (1986): "The Origins of Scientific 'Law'," *Journal of the History of Ideas* 47, 341–59.

Swartz, N. (1985): *The Concept of Physical Law.* New York: Cambridge University Press.

Swartz, N. (1995): "A Neo-Humean Perspective: Laws as Regularities," in F. Weinert (ed.),

Laws of Nature: Essays on the Philosophical, Scientific and Historical Dimensions. Berlin: de Gruyter.

Tooley, M. (1977): "The Nature of Law," *Canadian Journal of Philosophy* 7, 667–98.

Van Fraassen, B. C. (1989): *Laws and Symmetry*. New York: Oxford University Press.

Wertheim, M. (1997): *Pythagoras' Trousers: God, Physics and the Gender Wars*. London: Fourth Estate.

SELECTED BIBLIOGRAPHY

The following does not include entries for the readings in this collection. It spans the period from 1946 to 2002.

Books

Armstrong, D. *What Is a Law of Nature?* Cambridge: Cambridge University Press, 1983.

Bigelow, J., and Pargetter, R. *Science and Necessity.* Cambridge: Cambridge University Press, 1990.

Carroll, J. *Laws of Nature.* Cambridge: Cambridge University Press, 1994.

Cartwright, N. *How the Laws of Physics Lie.* Oxford: Clarendon Press, 1983.

Giere, R. *Science Without Laws.* Chicago: University of Chicago Press, 1999.

Goodman, N. *Fact, Fiction, and Forecast.* Cambridge: Harvard University Press, 1983.

Lange, M. *Natural Laws in Scientific Practice.* Oxford: Oxford University Press, 2000.

Lewis, D. *Counterfactuals.* Cambridge: Harvard University Press, 1973.

———. *Philosophical Papers, Volume II.* New York: Oxford University Press, 1986.

Nagel, E. *The Structure of Science.* London: Harcourt, Brace, and World, 1961.

Swartz, N. *The Concept of Physical Law.* Cambridge: Cambridge University Press, 1985.

Tooley, M. *Causation.* Oxford: Clarendon Press, 1987.

Van Fraassen, B. *Laws and Symmetry.* Oxford: Clarendon Press, 1989.

Essays and Articles

Armstrong, D. "What Makes Induction Rational?" *Dialogue* 30 (1991): 503–11.

———. "The Identification Problem and the Inference Problem." *Philosophy and Phenomenological Research* 53 (1993): 421–22.

Ayer, A. "What Is a Law of Nature?" *Revue Internationale de Philosophia* 10 (1956): 144–65.

Blackburn, S. "Morals and Modals." In *Fact, Science and Morality,* edited by G. Macdonald and C. Wright. Oxford: Basil Blackwell, 1986.

Carroll, J. "Ontology and the Laws of Nature." *Australasian Journal of Philosophy* 65 (1987): 261–76.

———. "The Humean Tradition." *Philosophical Review* 99 (1990): 185–219.

Cartwright, N. "Where Do the Laws of Nature Come From?" *Dialectica* 51 (1997): 65–78.

———. "In Favor of Laws that Are Not Ceteris Paribus After All." *Erkenntnis* 57 (2002): 425–39.

Chisholm, R. "The Contrary-to-Fact Conditional." *Mind* 55 (1946): 289–307.

———. "Law Statements and Counterfactual Inference." *Analysis* 15 (1955): 97–105.

Earman, J. "The Universality of Laws." *Philosophy of Science* 45 (1978): 173–81.

———. "Laws of Nature: The Empiricist Challenge." In *D. M. Armstrong,* edited by R. Bogdan. Dordrecht: D. Reidel Publishing Company, 1984.

Earman, J., Roberts, J., and Smith, S. "*Ceteris Paribus* Post." *Erkenntnis* 57 (2002): 281–301.

Fodor, J. "Making Mind Matter More." *Philosophical Topics* 17 (1989): 59–79.

———. "You Can Fool Some of the People All the Time, Everything Else Being Equal: Hedged Laws and Psychological Explanations." *Mind* 100 (1991): 19–34.

Foster, J. "Regularities, Laws of Nature, and the Existence of God." *Proceedings of the Aristotelian Society* 101 (2001): 145–62.

Goodman, N. "The Problem of Counterfactual Conditionals." *Journal of Philosophy* 44 (1947): 113–28.

Hempel, C. "Provisos: A Problem Concerning the Inferential Function of Scientific Laws." In *The Limits of Deductivism,* edited by A. Grunbaum and W. Salmon. Berkeley: University of California Press, 1988.

Hempel, C., and Oppenheim, P. "Studies in the Logic of Explanation." *Philosophy of Science* 15 (1948): 135–75.

Kneale, W. "Natural Laws and Contrary-to-Fact Conditionals." *Analysis* 10 (1950): 121–25.

———. "Universality and Necessity." *British Journal for the Philosophy of Science* 12 (1961): 89–102.

Lange, M. "Who's Afraid of Ceteris-Paribus Laws? Or: How I Learned to Stop Worrying and Love Them." *Erkenntnis* 57 (2002): 407–23.

Lewis, D. "New Work for a Theory of Universals." *Australasian Journal of Philosophy* 61 (1983): 343–77.

———. "Chance and Credence: Humean Supervenience Debugged." *Mind* 103 (1994): 473–90.

Loewer, B., and E. Lepore. "Mind Matters." *Journal of Philosophy* 84 (1987): 630–42.

———. "More on Making Mind Matter." *Philosophical Topics* 17 (1989): 175–91.

Lyon, A. "The Immutable Laws of Nature." *Proceedings of the Aristotelian Society* 77 (1977): 107–26.

McCall, S. "Time and the Physical Modalities." *Monist* 53 (1969): 426–46.

Menzies, P. "Laws of Nature, Modality and Humean Supervenience." In *Ontology,*

Causality and Mind, edited by J. Bacon, K. Campbell, and L. Reinhardt. Cambridge: Cambridge University Press, 1993.

Molnar, G. "Kneale's Argument Revisited." *Philosophical Review* 78 (1969): 79–89.

Musgrave, A. "Wittgensteinian Instrumentalism." *Theoria* 7 (1981): 65–105.

Pietroski, P., and G. Rey. "When Other Things Aren't Equal: Saving *Ceteris Paribus* Laws from Vacuity." *British Journal for the Philosophy of Science* 46 (1995): 81–110.

Pargetter, R. "Laws and Modal Realism." *Philosophical Studies* 46 (1984): 335–47.

Roberts, J. "Lewis, Carroll, and Seeing through the Looking Glass." *Australasian Journal of Philosophy* 76 (1998): 426–38.

Ruby, J. "The Origins of Scientific 'Law.'" *Journal of the History of Ideas* 47 (1986): 341–59.

Schiffer, S. "Ceteris Paribus Laws." *Mind* 100 (1991): 1–17.

Scriven, M. "The Key Property of Physical Laws—Inaccuracy." In *Current Issues in the Philosophy of Science,* edited by H. Feigl and G. Maxwell. New York: Holt, Rinehart, and Winston, 1961.

Sellars, W. "Concepts as Involving Laws and Inconceivable without Them." *Philosophy of Science* 15 (1948): 287–315.

Shalkowski, S. "Supervenience and Causal Necessity" *Synthese* 90 (1992): 55–87.

Shoemaker, S. "Causality and Properties." In *Time and Cause,* edited by P. van Inwagen. Dordrecht: D. Reidel Publishing Company, 1980.

———. "Causal and Metaphysical Necessity." *Pacific Philosophical Quarterly* 79 (1998): 59–77.

Sidelle, A. "On the Metaphysical Contingency of Laws of Nature." In *Conceivability and Possibility,* edited by T. Szabó Gendler and J. Hawthorne. Oxford: Clarendon Press, 2002.

Swoyer, C. "The Nature of Natural Laws." *Australasian Journal of Philosophy* 60 (1982): 203–23.

Vallentyne, P. "Explicating Lawhood." *Philosophy of Science* 55 (1988): 598–613.

Ward, B. "Humeanism without Humean Supervenience: A Projectivist Account of Laws and Possibilities." *Philosophical Studies* 107 (2002): 191–218.

Woodward, J. "Realism about Laws." *Erkenntnis* 36 (1992): 181–218.

———. "There Is No Such Thing as a *Ceteris-Paribus* Law." *Erkenntnis* 57 (2002): 303–28.

INDEX